OUR SOLAR SYSTEM FOR KIDS 8–12

The Young Astronaut's Guide to the Planets, Moons, Asteroids, & the Science of The Universe, with Space Facts & Adventures for Curious Minds

PANTHEON SPACE ACADEMY

Disclaimer Notice:
Please note that the information contained within this document is for educational and entertainment purposes only. Fun Facts Space Trivia took extraordinary research and studying effort to present accurate, up-to-date, and reliable, complete information. No warranties of any kind are declared or implied. Readers acknowledge that the author is not rendering legal, financial, medical, or professional advice. Pantheon publishers sorted the content within this book from various sources. Please consult a licensed professional before attempting any techniques outlined in this book.

By reading this document, the reader agrees that under no circumstances is the author responsible for any losses, direct or indirect, which could incur due to the use of the information contained within this document, including errors, omissions, and omissions or inaccuracies.

OUR SOLAR SYSTEM
FOR KIDS 8-12

CLAIM YOUR FREE SPACE EXPLORER ACTIVITY PACK!

Included with your book purchase

Your adventure doesn't stop with this book. I've created an **exclusive bonus pack** with **over 200 pages** of images, activities, and puzzles to help you **see it, do it, test it, and achieve it** as a Junior Space Explorer.

- **Visual Guide** (stunning images that bring each chapter to life)
- **The Space Explorer Activity Book** (puzzles, word searches, and more)
- **Coloring Book Pages** (30+ printable pages of planets, rockets, and more)
- **Solar System Quiz** (30 fun trivia questions for curious minds)

Get your hands on these bonuses to spark creativity and bring your book to life. You'll also find instructions on how to build a scaled model of the solar system.

Plus, celebrate your progress with an official certificate you can print and display — proof that you're a Junior Space Explorer! All bonus content is age-appropriate and approved by parents.

Start today: https://www.pantheonspace.com/solarkids

TABLE OF CONTENTS

Introduction 1

Cosmic Passport...5

(Instructions)

Chapter 1 9

Buckle Up, We're Going to Space!

Chapter 2 21

Unraveling Our Universe

Chapter 3 27

Cosmic Clockwork

Chapter 4 35

The Big (Bang) Picture

Chapter 5 41

Our Superstar: The Sun

Chapter 6 55

Terrestrial Terrains: The Inner Planets

Chapter 7 75

Gaseous Giants: The Outer Planets

Chapter 8 95

The Minor Members

Chapter 9 105

Starry, Starry Nights

Chapter 10 113

Astronomical Adventuring

Chapter 11 125

To Infinity and Beyond!

Conclusion 133

References..139

INTRODUCTION

"What's that, mommy? Why is it moving?"

It's 9:45 p.m. As you get ready for bed, you glance out of your windows before closing the curtains, but then you catch a glimpse of a bright glow in the sky passing overhead. Is it a plane? Is it a satellite? Is it a meteorite? You aren't sure, but the mystery draws you in. Following the long history of generations before, you find yourself wondering what lies beyond the comforts of Earth.

Lately, you might feel like you are "going through a phase," becoming increasingly obsessed with and inspired by space. Just watching a sci-fi film isn't going to satisfy that itch to know more. Perhaps you aren't sure what to make of this newfound interest. After all, many people wonder about life in space, dreaming of visiting a space station or walking on the moon. So, you're asking yourself, "How do I get there? What is it like to launch into space? How dangerous could all that emptiness be?".

Our Solar System for Kids offers you and your family the tools to interact with and learn more about space in a meaningful way. Through this enjoyable learning experience, you will spend quality time together as you navigate a new hobby, add to your enthusiast's collection, and gain perspective on yourself and nature. Adventurers of all ages are sure to enjoy the fun yet informative narrative. Space is a great teacher—a world that commands respect and inspires imagination.

Our Mission

Pantheon Space Academy, a group of astronomy enthusiasts, believes that the best form of self-confidence and self-esteem comes from pursuing knowledge and achieving goals. With expertise in science, teaching, and storytelling, we at Pantheon Space Academy are passionate about all things space and strive to make space exploration accessible and engaging for every family. Many members of Pantheon Space have been studying astronomy since childhood. We study planets, constellations, and other space phenomena collectively, working to better understand and share our increasing knowledge of space. As parents, teachers, and students, we are committed to quality education that optimizes both retention and engagement.

Today, our understanding of astronomy and space is more significant than ever, yet many unknown details await discovery. At the same time, the mystery and awe of space make learning more exciting. You could say that space is a constant adventure. Pantheon Space worked tirelessly to make this journey as eventful and informative as possible. Offering an opportunity to build self-esteem, inspire imagination, and increase knowledge, Pantheon Space hopes to support your path to academic achievement. Along the way, you will learn a great deal about our place in the universe, have fun, and spend quality time with your family.

An Overview

This book, *Our Solar System for Kids*, will hold many answers to your questions about space. In Chapter 1, the basics of space suits, the space shuttle, and escape velocity are discussed. After liftoff, the first stop in Chapter 2 will be an overview of the solar system, reviewing the different parts and placement of the planets and the Sun. Chapter 3 explains the mechanics of what holds the solar system together and introduces the minor members of the solar system. Then, the focus shifts to the Big Bang in Chapter 4 and how our solar system formed.

Chapter 5 focuses on the Sun, with Chapters 6 and 7 detailing the inner and outer planets, respectively. These chapters describe the planets' size, characteristics, and famous space landmarks—Chapter 8 details the minor members, including Pluto and the Kuiper Belt.

Focusing outward, Chapter 9 shifts to looking at the life cycle of a star and introducing the Milky Way concept. Chapter 10 explores the different ways that humans have explored the solar system and beyond. Finally, Chapter 11 details the characteristics of galaxies and their important place in the universe.

With a strong focus on precise information, solid facts, and entertaining descriptions, this book will inspire, excite, and educate. It's time to grab your cosmic passport and start a whirlwind tour of the universe!

Cosmic Passport

(Instructions)

Welcome, Astronaut!

Your space journey begins now. As you read *Our Solar System for Kids*, you'll visit every planet in our solar system — from the fiery Sun to the icy edges of Neptune.

Print your **Planet Cutout Sheet** from www.PantheonSpace.com/solarkids

Cut out each planet as you encounter it in the story and glue it onto your **Cosmic Passport** (the next page).

Each planet you collect is a new mission completed! When all nine missions are done, print your **Explorer Certificate** to celebrate your success.

Mission Goal:

Complete your passport to prove you've officially traveled the entire solar system.

Pro Tip:

You can even add small stickers or draw stars each time you learn something new about a planet!

COSMIC
PASSPORT

Junior Space Explorer Program – Mission Log SE 21

Pantheon Space Academy | #cosmicpassport

CHAPTER 1

BUCKLE UP, WE'RE GOING TO SPACE!

This is the waiting area. "I can't wait to see the Moon!" Please be careful with all your hopping up and down. For now, hold onto your mom's hand. "Are we going soon?" The shuttle van will be here any minute. Hi, young man. I notice that you aren't jumping up and down. You contain your excitement quite well. Are you nervous? Well, don't be because I'm going to be with you every step of the way. Look, the shuttle van is here and will drive you to the launch pad.

"Will we be able to visit Saturn? I'm excited to see Saturn's rings." Yes, and we will see her moons. Look out your window. Can you see in the distance? That's the upright silhouette of our spaceship. You will buckle in and rocket to outer space in a few hours.

Hi everybody, gather around. Welcome to Orion Tours. Today, you will officially become a space traveler. Please finish checking your backpack and all your luggage bags. The Dragon's Lance and its passengers will visit several Orion observatories stationed around the solar system over the next few weeks. You will have firsthand experience of life on many of our solar system's planets.

T-minus three hours.

Three hours! It's getting close to launch! Soon you'll be leaving planet Earth and heading toward the stars!

Before that, the main command center and crew must double-check everything to ensure that all systems are operational. Since space travel is so expensive and dangerous, many details must be checked and double-checked before liftoff. These preparations take a long time, and there is a strict schedule to follow. The countdown must keep on target because the mission's launch window doesn't stay open forever.

The launch window describes a set block of time when the space shuttle can lift off. Depending on where you want to go, this window can be very short—only a few minutes long, or very long, like a month. Sometimes, bad weather can limit or cancel a launch window. This is why it's essential to be very organized and plan as best as you can for the space shuttle launch.

When NASA analyzes its goals and the cycles of our solar system, they can target a specific launch window and begin preparations. When they are almost two days away from launch, the shuttle test director activates the countdown clock, and authorized personnel are activated to double-check the space shuttle systems. They want to ensure that the computers, the navigational systems, and the power systems are working so that the shuttle can communicate and fly well.

About one day before liftoff, workers who aren't necessary for launch leave the pad as the space shuttle receives fuel. It's just like how your parents have to fill your car with gas so they can drive you to school. The space shuttle also needs fuel—a lot of fuel. Engineers must recheck other important computers and systems to ensure communication between the shuttle and Earth will work.

T-minus two hours.

Now it's time to get into the van and drive to the launch site. You will join the pilot and the flight crew in the White Room and prepare for launch! The cleaning of this room is very particular, intentionally designed and built to stop dirt, dust, or stray hair from getting into the space shuttle. This way, the shuttle can operate smoothly while in space. Enter in, and we will put on our spacesuits.

For anyone leaving Earth, a space suit is an essential protection, especially when entering and exiting orbit. Spacesuits have many different parts, and each unique piece will help keep you alive in space. This is because outer space holds many dangers: heat, cold, and radiation are extremely dangerous and nothing like what we experience here on Earth. Preparing for super high and super low temperatures isn't too tricky. Still, when the Sun's or other stars' rays hit you with particles, you can experience harmful radiation to your body. Small dust and rocks that are moving fast can also hit you and hurt you badly. On top of that, space doesn't have a lot of oxygen, so you would blackout in 15 seconds without an oxygen tank.

At the beginning of space exploration, suits were not very protective, and you couldn't go on spacewalks in them. Scientists and engineers are still improving upon the Extravehicular Mobility Unit (EMU) and other spacesuits. The EMU suit is made of soft and hard materials designed to protect astronauts and allow them to move around more easily. With roughly 13 layers, you can imagine this suit is very complicated and expensive! However, you will only wear the bulky EMU suits when you leave the shuttle, but we can orientate you on the parts of the EMU suit, so you will be prepared when the time comes to put them on.

Let's start with the helmet. The bike helmet you wear when biking or roller-blading protects your head during accidents. In the same way, the space suit's helmet protects your eyes and skull from the Sun's rays. The helmet visor, covered with a gold coating, allows you to see without getting harmed by radiation or the strong sunlight. There are unique ventilation systems designed to help you breathe in the helmet!

Next are the arms and gloves. The center of your body might keep warm in space, but your hands can freeze. These gloves are unique because they have built-in heaters to keep your fingers warm! They also allow you to move your hands efficiently to use tools while working in space.

Made of hard fiberglass (similar to materials that boats are made of nowadays), the HUT covers your arms and chest down to your waist. It connects to your gloves and Lower Torso Assembly. Attached to the HUT are two vital systems: the Portable Life Support System and the Displays and Control Module, aka DCM. Here you can control the suit's air pressure, airflow, and temperature.

Now, these are the space pants! They are called the Lower Torso Assembly. Like the rest of your suit, this part protects your body from the harsh world of space, but it is also advantageous. Around the waist, you'll find attached rings so that you can tie yourself to the space station while working outside. You can also hook different tools to the "belt" during a spacewalk!

The Portable Life Support System is your super special backpack attached to the Hard Upper Torso. It holds many relevant systems for the astronaut, including an electric battery, a water tank, a two-way radio, a fan, and a carbon dioxide removal system. Without the PLSS, nobody would be able to survive in space.

It can get scorching in space, so one of the first layers to be put on is a critical long underwear-like piece of clothing called the Liquid

Cooling and Ventilation Garment. With small tubing running through the material, the LCVG pushes cool water from the PLSS all around the body to keep astronauts cool. This way, space travelers won't get overheated from the Sun!

There are many other necessary pieces to the spacesuit, like a drink bag that the astronaut can use to stay hydrated or the communication devices that help the astronaut contact the space station, the space shuttle, and NASA. We can use these in case of an emergency.

Knowing how integral the suit is, be very careful when putting it on. Each spacesuit costs millions of dollars! When we are safely inside the spacecraft, you know you can wear the more comfortable, sleek space suits designed by SpaceX. These suits are also expensive. More than just clothes, they will protect us as we exit and enter the orbit of Earth and Mars, providing us with oxygen and proper air pressure in case of accidents.

Now that you have on your spacesuit, it's time to look over the shuttle!

T-minus one hour.

The previous space shuttle built, the Endeavour, was decommissioned with the older space shuttle Atlantis in 2011. When NASA terminated its space program, one reason was how expensive the space shuttles were—around 1.7 billion dollars each! Although the solid rocket boosters and orbiters were recyclable, the external tank, which held most of the fuel, always burned up in the atmosphere. Since then, alternative companies like SpaceX and Boeing have been developing alternatives to the space shuttle. In

2050, Orion Tours will fly with the Dragon III, often called Dragon's Lance. Follow me, please, let's take a look at the three main parts of this futuristic space shuttle.

The Dragon's Lance has two long rockets called the Solid Rocket Boosters. On either side of the ship's main body, these rocket boosters might look smaller, but they hold great power! The boosters carry solid fuel that burns first, helping push the core stage rockets and the Dragon's Lance vehicle upward. Each rocket booster holds one million pounds of specially mixed solid fuel. Endeavour's old solid rocket boosters were recyclable. After two minutes of burning around five million pounds of thrust, the two rocket boosters fall off and parachute into the ocean to be picked up and used again. Today, these solid rocket boosters work more efficiently, but they are not reusable.

On top of the long, thin body of the primary core rockets and modules, the crew vehicle sits. That's the tip of the Dragon's Lance— it's where everybody will stay as the tour travels from place to place. The Endeavour used a separate shuttle that could fly and glide back to a runway. This capsule ship will fall into the ocean instead with the help of parachutes. In a few weeks, the Dragon's Lance will splash down off the coast of Florida. That's still a long way off, though.

The Endeavour used to have three parts: the forward, mid, and aft fuselage. The crew living quarters were located in the forward section, while the mid-fuselage held the payload bay. The payload bay has a tremendous amount of storage space. The Endeavour carried satellites, equipment for the International Space Station, or even a temporary, air-pressurized room for a lab. Further down at the bottom of Endeavour was the aft fuselage, where all of the big engines were housed. The RS-25 engines were powered by liquid hydrogen and oxygen stored in the external tank. Once the external tank fell back to Earth, engineers could no longer use these engines. Instead, the Orbital Maneuvering System rockets and thrusters guided the Endeavour.

Nowadays, the main spacecraft comprises many parts of rockets, each one used at different stages. Some are reusable, and some are not—depending on the design. Today, we will be flying with the Dragon's Lance, based on the sleeker and efficient layouts of the 21st century. The crew vehicle, located toward the top of the long, thin spacecraft, is where we will stay for the most part. It includes a state-of-the-art Observatory module with 360-degree views! Below us, other modules and sections of the rocket will boost us into orbit and beyond to the inner solar system planets. We will burn another installment to get to Mars, which is why Dragon's Lance is so big. Above Mars, a specialized staging area will add sections for final boosts to the outer solar system regions.

Overall, the Dragon's Lance is a modern engineering and technology feat. The Orion Tour engineers have cleared us for departure. Are you — *T-minus two minutes.*

Oops, I'll explain later. That's our cue to board. You are about to experience your first liftoff! I saved you the best seat. Sit here behind the commander and pilot. You can see that preparing for liftoff takes oodles of mathematical and scientific calculations. Many checklists have to be tended to and marked off. Engineers must recheck the weather, launch window, state of the space shuttle, and even the area around the launch pad. The Control tower will check all of the computers, technology, and communication devices a few times. They work really hard to ensure our safety.

What you're hearing now is the automatic ground launch sequencer start. Different supports for the rocket are retracting. The Orbiter cuts off connection to ground power and now relies only on its internal energy.

The main engine is now starting!

3 . . . 2 . . . 1 . . .

Light the candle! Everything you're feeling and seeing is normal. As the ship roars upward, the pressure pushes you back in your seat, but you shouldn't be scared. You're on the best ride ever! Now, at two minutes, you'll see the two solid rocket boosters detach. We are tearing through the sky toward outer space. Do you see how the monitor tells you which part of the Earth's atmosphere you are passing through? Outer space is getting close!

How close is outer space? Outer space is only around 60 miles above the Earth. What's happening? I know it's hard to move your head right now as our ship is rocketing through Earth's protective atmospheric layers. You can't see outside right now because our helmets are on, and you are strapped into your seats, but you can stare up at the computer screen in front of you.

Is the spacecraft on fire? No, it's okay. The blur of heat waves and smoke is normal. Whether you are exiting or entering Earth's atmosphere, the high speeds and friction against the atmosphere particles cause intense heat. You can check the computer to see the position of the Dragon's Lance.

You will notice that the rocket has already left the troposphere. That is the layer we live in, with very dense particles of oxygen for breathing. Closer to the poles, the troposphere is around five miles thick but rises to 10 miles at the equator. Since this layer is closest to Earth, it has more dust, ash particles, and water vapor, forming clouds. Now we are over the clouds, and the rocket is entering the stratosphere.

Here, the air is thinner, but you begin to feel warmer—a sign that you are hitting the stratosphere, where you can find the ozone layer. This layer of molecules protects the Earth from harmful UV rays and warms up as it receives the Sun's radiation. Only large commercial airplanes fly at this altitude, so the sky feels pretty empty. That was incredibly fast. Dragon's Lance is now 31 miles above the Earth, where the stratosphere ends.

After the warm band of the stratosphere, things start to get chilly again in the mesosphere. It's virtually empty up here because regular planes can't fly at this altitude. Sometimes, you might see the occasional military jet, but otherwise, it's pretty quiet in the mesosphere. At night, when you see a streak of light shooting to Earth, those are meteorites burning up in the mesosphere. Even though the mesosphere isn't as thick as the lower layers, it is still excellent protection against meteors! Around 50 miles away from Earth, the mesosphere ends. Outer space is very close now!

We are now entering the thermosphere! The start of space is the invisible Karman Line, located 62 miles above sea level. Here, the temperatures begin to increase again because of the Sun's radiation, sometimes reaching 3,600 °F! In this area, you will only see spacecraft moved by rockets and thrusters: the International Space Station and the space shuttle. Look down at Earth with me. We see a ball of swirling blue and white and a distant glimpse of continents. At around 440 miles, the thermosphere ends. Are you ready to explore what lies beyond?

We can now celebrate as the Dragon's Lance reaches orbit. Watch as sections of the lower rockets fall away. Now, you are floating in the zero-gravity world of the thermosphere! Look opposite of Earth to see what lies beyond the thermosphere: the exosphere. Between 440 and 6,200 miles, the exosphere of the Earth holds the last few particles of gas and oxygen. With even less gravity here, the particles drift off into space. The only man-made aircraft hanging out here would be various satellites.

This is when I like to look back at Earth. I realize how much I love home and that flying on the Dragon's Lance is super special. Less than 600 people have been to space, and you can mark today as the trip of a lifetime!

CHAPTER 2

UNRAVELING OUR UNIVERSE

I f you floated away from the Dragon's Lance and looked around, what would you see? If you answered, "the solar system," you are correct! The neighborhood of our Earth is called the solar system. Our Solar System includes one sun and eight planets, and I can't wait for us to visit all of them! For now, let's look at the layout of our solar neighborhood, so we can understand where each planet is and how big or small they are compared to the Sun.

We look over the solar system and make a few observations immediately. For one thing, you can see that the Sun is truly the star of the solar system, and it is the most prominent member of our solar system, so its gravity is responsible for keeping us all together!

Also, you notice that all the planets move in differently shaped circles around the Sun. Some of the spheres move faster, while some move slower. Some planets orbit the Sun at the same distance, but other system members might move very closely and then travel far away from the Sun.

As you look closely at the first four planets closest to the Sun, you realize they are smaller and have solid terrain. However, the other four planets—the ones furthest from the Sun—are giant and have no solid ground to stand on. They are called gas giants. Your cosmic passport allows you to visit them all! Still, journeying around the solar system is going to take time. After all, the Sun and its planets are all very far apart. Thanks to the Dragon's Lance and its rockets, you will be able to visit most of the solar system in the next few weeks!

If we were to make a real-life scale model of our solar system, we would have to go outside to build it. In fact, we can use a professional football field to show how far away each solar system member is. To represent the Sun, we will set a 12-inch basketball on the ground. A basketball doesn't seem colossal, but the Sun is, in fact, very massive. Over one million Earths could fill up our Sun!

The Sun is a great place to start, but there is so much more to the solar system!

The first planet of the inner solar system, Mercury, stands closest to the Sun. It is tiny. Put a pinhead down at 12 yards away to show how small it is. Wow: A pinhead is tiny. That is because Mercury isn't a big planet—it's about one-third of the size of Earth. Is the next planet bulkier?

Well, kind of. Compared to the basketball Sun, Venus is only the size of a quilting pinhead. It isn't super far at 22 yards away from the Sun, so you can only imagine how hot Venus must be!

Earth is almost the same size as Venus, so it's also the size of a quilting pin. Fortunately, Earth is 31 yards away from the basketball Sun, which helps lessen the impact of the Sun's rays while still providing warmth.

What about Mars? Mars is the last planet of the inner solar system, and it's about half the size of Earth, so a pinhead can also represent it. If you place it 47 yards away from the basketball Sun, you can see that Mars looks very small and very far away from the Sun. It isn't the last planet, though!

Beyond Mars, the asteroid belt hangs. The belt wraps around the Sun, creating a barrier between Mars and the gas giants. Asteroids are so minor that you would use dust or tiny glitter to represent them. If we look past the asteroid field, we can see that Jupiter, Saturn,

Uranus, and Neptune are very far away from the Sun. Although they are enormous, none compare in size to the Sun!

Jupiter is the most massive planet. It's about 11 times the size of Earth, so you can use a ping-pong ball to represent this great gas giant. How far away is Jupiter from the basketball Sun? It's 162 yards! That's so far away; if the basketball Sun sat on a football goalpost, the ping-pong ball Jupiter would be in the stands on the other end of the stadium!

Of course, Saturn is even further. Saturn is just a bit smaller than Jupiter and is only nine times the size of Earth. You could use a one-inch ball bearing to represent the ringed planet of Saturn. At 297 yards, Saturn is also in the stadium seats.

You can only imagine how far away Uranus and Neptune might be. These two are around the same size, about four times the size of Earth. They are about half the size of Saturn, so you could use half-inch ball bearings to represent them both. However, Uranus is one-third of a mile away from the basketball sun, and Neptune is half a mile away! That is pretty much down the street! At these distances, the Sun must appear very small. Living here on the edge of space would be very cold!

The nature of space is so vast; it can be hard to understand how far away everything is. We can catch a glimmer if we look far into the distance where the closest star is. In our real-world model, the nearest basketball star to our solar system would be 4,600 miles away. That's further than the distance between New York and the North Pole! That's pretty mind-blowing. Right?

With so much space to explore, where can we start? How about we learn some more details about our space neighborhood? After all, there are many other places and members of the solar system to visit. Plus, we can find out the best way to measure space distances. Let's get back on the Dragon's Lance.

We'll need a good rest before we get back to exploring the solar system.

CHAPTER 3

COSMIC CLOCKWORK

Earlier, when we looked over the solar system, we noticed that the planets were moving around the Sun. These are called orbits. Looking closer at the details of the solar system, you begin to realize that the planets aren't the only things drawn to the Sun. There are other members of our space neighborhood that fly around the Sun as well. You may have wondered, "Why are all of these planets and objects moving around like that? Why aren't they just wandering off into space?"

The truth is that the Sun holds a lot of power. Not only is it scorching, warming the Earth, and giving us our daily cycle, but it also has a strong gravitational pull that keeps everything in orbit. Just like gravity on Earth keeps us walking on the ground, in the same way, the Sun also has a gravitational pull that keeps the Earth and all of the other members of the solar system in orbit. I hope that helps you realize just how powerful the Sun is.

One of the fundamental factors for gravity is mass. Objects with more mass have more gravity. Gravity is also affected by distance. The closer you are to a stable gravitational mass, the more gravity you will feel. For example, the center of the Earth has a strong gravitational pull due to the molten core underneath the Earth's crust of rock and soil. Thanks to Earth's gravity, we can walk around without falling or floating off into space. We can also measure our weight based on the effect of gravity on our mass.

However, the mass of the Sun is far greater than Earth's, so its gravitational pull is way more robust. If the Sun were not as hot and had firm ground to walk on, our weight on the Sun would break the scales! Thanks to the Sun's dense mass, its gravitational pull attracts the planets, comets, asteroids, and other dwarf planets in orbit. Earth's steady orbit in the sweet spot of the solar system allows it to have superb amounts of warmth without getting too hot or too cold for life.

There are many ways to look at and describe an orbit, whether you are talking about the Moon orbiting the Earth, the Earth orbiting the Sun, or the solar system orbiting the center of our Milky Way galaxy.

For example, not all planets rotate the same way! If an orbiting body moves in the same direction as the planet's rotation, then the orbit is in a prograde direction. However, if the orbit is opposite to the planet's rotation, then the member is in retrograde. This phenomenon affects how it experiences solar and sidereal days.

On top of that, not all planets move around the Sun in perfect circles. Many solar system members move closer to and then shift further away from the Sun during their orbits. When the planets come closer to the Sun, they enter the perihelion of their orbit. As they move away, they enter the aphelion of their orbit. On Earth, we experience its aphelion when it's summer in the Northern Hemisphere.

Orbits impact how we view time as well. As planets orbit the Sun, they also spin on their axis. One rotation is considered a day. However, there are two kinds of days: solar and sidereal days. Solar days measure the time it takes for a point on a planet facing the Sun to come around and face the Sun again. Sidereal days focus instead on stars outside of the solar system. Due to rotation and revolution cycles, solar days are longer than their sidereal days for most planets.

However, planets with retrograde rotation have a longer sidereal day than solar.

Not all planets orbit at a 90° angle, similar to how the Moon's orbit follows the Earth's equator. Almost all of the planets orbit close to the same plane, but a few planets circle differently. For example, Mercury has a slight tilt to its orbit at a 7° angle, and the dwarf planet Pluto has a 17° angled orbit around the Sun!

Whether we are sending ships to Mars or probes to the solar system's outer reaches, understanding orbits helps us plan space travel more efficiently. For example, as we prepare for future trips to Mars, we can shorten travel time by choosing a launch window where Mars and Earth are closer together. Calculating orbits also helps us know when and where to look for traveling celestial bodies like comets or dwarf planets like Pluto.

Speaking of Pluto, planets aren't the only thing you find in our solar system. With the sensors on the Dragon's Lance, you will begin to notice some fascinating space objects that are smaller and less regular than planets: moons, asteroids, dwarf planets, and comets.

Come gather around the control center; let's look closely at the moons in the solar system. Most nights of the month, if you look out of your home's windows, you will see Earth's Moon shining in the sky. Imagine looking out of your window and seeing more than one! Although Mercury and Venus don't have any moons, many other planets have multiple moons. Mars, for example, has two moons, while Jupiter and Saturn have over 50 moons each! Some of the moons in our solar system, such as Jupiter's Europa and Saturn's Titan, provide us with mysteries we have not yet solved.

Follow me to this window. Do you notice the rocks floating around? You know what they are! Asteroids. In every shape and size, they can be found everywhere in our solar system. There are around one million, but most of them are found in the asteroid belt between Mars and Jupiter. Some asteroids are hefty, almost 329 miles in diameter. Others are only 33 feet across. Most are a combination of clay, nickel-iron, and silicate rock.

Of course, there are the compelling dwarf planets: Pluto, Eris, Ceres, Makemake, and Haumea. Although they orbit the Sun like the other planets, dwarf planets aren't quite spherical and orbit closely to other space objects. Pluto has "moons" that orbit it. However, all of these dwarf planets are smaller than Earth's moon, which explains their name and category.

Whoa! That's a comet we just caught a glimpse of. Thanks to their rarity, comets have always been noticed throughout ancient times, which meant that ancient astronomers thought they were magical signs. Although we rarely see comets while on Earth, we have a better chance of finding them in our solar system now that we're in outer space. So far, we have counted 3,736 comets. Thanks to their often irregular and slow orbits, we have to rely on probes and telescopes to track their positions.

Where do comets come from? Many astronomers believe that comets are formed and ejected from the Kuiper Belt or further beyond in the Oort Cloud. The Kuiper Belt is closer, found just beyond the planet Neptune. Like the asteroid belt, the Kuiper Belt is full of rubble, but it is essentially frozen and includes most dwarf planets. The Oort Cloud is further yet. Scientists believe it surrounds the entire solar system. A large sphere of frozen space rubble ranging from small space dust to large mountains of frozen debris. The Oort Cloud forms a shield that protects the solar system with a rocky buffer. Understanding what the Oort Cloud is made of exactly is difficult, since it is so far away, we haven't yet been able to reach it. This just goes to show that when it comes to space exploration, distance absolutely matters!

Talking about the distance of planets, the orbital distances, and the far-off Oort Cloud might be causing you to wonder how far away everything is. There are two ways to measure distance on Earth today: imperial (miles) and metric (kilometers). However, using big numbers all the time makes calculations quite hard!

For example, four quadrillions is the distance between us and a star that astronomers have discovered in the Milky Way galaxy. That's bigger than five commas! Trillion is four commas, billion is three commas, and million is just two. Don't worry about that for now. You can see that this number is way too big to use for easy calculations, especially since our universe is so big. Thankfully, astronomers created a way to measure distance in space.

When describing distance in our solar system, we will use AU units. AU means "astronomical unit." One astronomical unit equals ninety-two million nine hundred fifty-five thousand eight hundred and seven miles, to be exact. Since orbits can increase or decrease the distance of a planet from the Sun, we use an average to describe how far away a world is. Earth is one AU away from the Sun, while Neptune is 30 AU. Other minor members of the solar system tend to be very far away from the Sun. For example, Pluto is around 68 AU away from the Sun. The Kuiper Belt is even further off—approximately 30 to 55 AU. Can you guess how distant the Oort Cloud is? That's right, 5,000 AU! That's unquestionably far. If we drove there in a car at 60 miles per hour, it would take us around 885,000 years to get there!

Do you know the distance to the closest star? As unbelievable as it sounds, the distance to the closest solar system is so huge that scientists require another form of distance measurement: light-years. Let's consider the mysterious star known as Betelgeuse. This star is so far away; we have to use mathematical formulas, like exponents, to express the number easily in miles or even astronomical units! Betelgeuse is four quadrillion miles away. That's about 45 million AU.

As you can see, multiplying or using these numbers would be very hard. You could use exponents to make these numbers a bit shorter, but scientists decided on a new way to express distance. Thanks to light, we can calculate a new form of speed and distance.

Light moves constantly throughout our universe. It can be seen from very far away. This is because light moves very quickly through space. Using the theory of special relativity, scientists have discovered that light travels at 670,616,629 miles per hour. That sounds pretty fast, right? Think about it this way: If you could travel at the speed of light, you could go around the Earth 7 ½ times in ONE second.

With this new way of measuring distance, we can rethink how to express the distance between the Sun and a star like Betelgeuse. Although far-off stars are difficult to calculate, most scientists agree that Betelgeuse is around 724 light-years away. Using 724 LY is much easier to use as a measurement than 4 quadrillion 256 trillion 124 billion 891 million miles! However, sometimes the modest numbers used for light-years don't help us appreciate the vast distances of our universe.

Later, when we talk about our Sun and its nearest neighbor, we can use light-years to talk about the distances more swiftly. Let's now learn how our Milky Way and solar system came to be. Hint: It didn't start with a massive explosion!

CHAPTER 4

THE BIG (BANG) PICTURE

What do you think of when you hear the words "big bang"? Do you imagine an explosion? Maybe you see a bright flash of light quickly spreading, followed by a very loud sound. The beginning of the universe sounds pretty exciting, doesn't it?

Today, scientists don't think the Big Bang involved a large explosion since the universe's initial expansion happened too quickly. According to most current theories, a very dense and hot point in space suddenly expanded within less than a second into the early state of the universe.

After another microscopic period of time, less than a second, the universe grew again. Many scientists call this "the inflation" because the universe kept growing, doubling in size around 90 times. As it grew, it got less hot and less dense. There wasn't much focused light at this time, and there weren't any galaxies, solar systems, or even planets. Then, 400,000 years passed before atoms started to form. As the dense clumps of gas began to collapse, the first galaxies and stars were created.

I know you have so many questions about our home galaxy. According to many scientists, the Milky Way galaxy likely formed when the other galaxies were created—13.6 BILLION years ago! Around nine billion years after the Big Bang, our solar system began to form. However, our Sun and planets didn't look the same as they do today.

If we traveled back in time 4.5 billion years ago, we would see an immense cloud of gas and dust, which we can call "the solar nebula." There are different theories as to what happened next. Did the Milky Way begin to cool down too quickly for the nebula? Or did a nearby supernova emit a shockwave? Either way, the gas began to get denser. It clumped together until gravity caused the cloud to collapse, and our solar system began to take form in its first strange shape.

When the solar nebula collapsed, it didn't suddenly make a Sun and planets. Instead, it began to spin around and flatten into a disk shape, like a Frisbee. The cloud of gas and dust moved toward the center of the disk, where a young protostar was starting to form. Was the disk going to stay that way forever? Not if gravity had its way!

Thanks to gravity, more and more material was pulled into the protostar, increasing the pressure so much that hydrogen atoms combined and formed helium. This chemical reaction caused the Sun to form as it released tons of energy. The shockwave pushed all the other material back, so no further dust and gas fell into the Sun.

With the Sun just born, the rest of the solar system was still a bit of a mess. It took another 100 million years for the planets to form. That might seem like a long time, but compared to the whole life cycle of the universe, this was a tiny window of opportunity. There are many ideas about how the planets might have formed, including the core accretion model, the pebble accretion model, and the disk instability theory.

The most popular theory, the core accretion model, suggests that pieces of rock were flung about, crashing into each other during this time. Gas began to collect together as well. Some of the rock parts started to grow in size, creating small celestial bodies that we often call "planetesimals." Planetesimals are kind of like baby planets. Some of them attracted more mass and matter and became the planets we know today; others were destroyed. Some were flung off into the Kuiper Belt and the Oort Cloud, while the rest may have become moons for other planets in our solar system.

Not all planetesimals were the same, however. Today, we can see that the ones closer to the warmth of the Sun were made of metals, and other planetesimals further away were formed from ice. Some planetesimals attracted gas clouds, but their inner cores melted from the heat and pressure, creating the gas giants we appreciate today.

Although the core accretion model can't fully explain how the gas giants were created in such a short period of time, many scientists consider the theory reasonably sound to explain how the inner solar system was formed. Perhaps a combination of the core accretion model and the disk instability model can explain the presence of gas giants early on in our system. However, although we may have some good ideas about how planets formed, it's harder to understand how they got to their current-day positions and orbits.

When exoplanets were discovered in the early 2000s, scientists realized that massive objects in space could move around. They might be pushed or pulled by gravity or affected by other nearby masses or shockwaves. In particular, the gas giants might have been closer to the Sun at one point, but as the smaller, rockier planets were formed, they pushed the gas giants further away. This resulted in the gas giants losing their molten cores and keeping their gas.

Jupiter and Saturn were so mighty that they may have helped form the outer solar system, pushing Uranus and Neptune into more distant orbits and removing several planetesimals from the system.

Jupiter's gravity could have kicked out other gas giants, dwarf planets, or planetesimals from the system. As the inner and outer solar system planets settled into their orbits and rotations, the remaining planetesimals broke apart into the asteroid belt or disappeared into the Kuiper Belt and Oort Cloud. Others formed meteors, comets, irregular moons, or dwarf planets, like Pluto.

Since Earth fully formed and began to spawn life, the solar system has remained, for the most part, relatively stable. All eight planets have rotated on their axis and orbited the Sun for millions, if not billions of years. The inner four planets, formed from metals and silicates, circle the Sun more closely. Further away, the gas giants orbit the Sun, protecting us with their massive gravity pull by attracting larger, incoming planetesimals.

Although our neighborhood lies kind of close to us, the solar system still holds many mysteries. As we continue to search for life in the universe, the first areas we hope to explore are within our own backyard. For example, Mars and Europa (one of Jupiter's moons) offer mysterious glimpses into the possibility of human settlements on other planets or moons. For now, scientists observe and analyze the data we find from telescopes, rovers, and probes, learning about and appreciating even more of the terrific balance of our solar system.

Because the solar system has become stable and regular with its cycles and orbits, the universe has also stabilized. However, close to 100 years ago, Edwin Hubble looked through his telescope and noticed that the universe wasn't staying the same—it was expanding. Years later, other astronomers used the Hubble Space Telescope to study the stars, and they also noticed that the universe wasn't just expanding normally but amazingly growing faster and faster!

Why the universe hasn't stopped expanding is still a mystery, and some scientists believe it is due to the as-yet-unknown dark matter. Will the universe continue to stretch? Will it collapse inward, or will it go out with a big bang?

The question is still up for debate.

CHAPTER 5

OUR SUPERSTAR: THE SUN

The Dragons Lance has us in range to check out the Sun. We have to be very careful during our approach to the Sun. The engineers have given us clear instructions to keep our distance. Maybe we can get closer in the future with superior technology and spacecraft. Visiting the Sun would be super cool! Well, it would honestly be super hot! As the most prominent and dazzling member of our solar system, the Sun is the superstar of the show. Without it, life would never have happened, but the Sun also holds many dangers.

Even in ancient times, when astronomy was very basic, humans understood that the Sun meant both life and death. A famous Greek story tells of a man called Daedalus and his son, Icarus, who created wings to fly. When they were imprisoned, Daedalus and Icarus escaped by flying. However, Icarus ignored his father's warnings and flew too close to the Sun. The wax melted, destroying his wings and causing Icarus to fall into the sea. Do you think Icarus would have flown so high if he had known more about the Sun? Maybe Icarus couldn't help but want to fly higher, see further, and learn more. Today, thanks to science and technology, we can lower the dangers of studying the Sun and help us understand ways to help protect Earth from its clear and present power.

To us, the Sun seems very big, and its heat and energy are unmatched by any other member of the solar system. However, compared to other stars in the Milky Way, the Sun is not that big, which means that our Earth's atmosphere is appropriately heated and can support life. If the Sun had been too big or too small, Earth would be too hot or too cold. Here are some basic facts about the Sun.

- **Star Type: Yellow Dwarf**
- **Distance From Earth: 1 AU (92.92 million miles)**
- **Radius: 432,168.8 miles**
- **Rotation (a Day): 25-36 Earth days**
- **Orbit Around the Milky Way: 230 million years**

Wait. The Sun is orbiting around the Milky Way's center? It's hard to imagine, but the Sun has its own orbit around the center of the Milky Way galaxy. The Sun and the solar system are, in fact, speeding along at 450,000 miles an hour!

That's not the only thing crazy about the Sun. As the Dragon's Lance draws closer to Venus and Mercury, the inside of the spacecraft begins to feel like a sauna. I think we can all agree not to get much closer to the Sun. Walking on it is impossible anyways. For one thing, the Sun doesn't have ground as the Earth does. Plus, the ship would burn up if it got too close. This is because the atmosphere of the Sun is just as hot as some of its inner layers. Some of the layers have never been seen by the human eye. Buried deep down under layers of heat, gases, and radioactive matter, the center of the Sun remains a mystery.

We're approaching our first stopping point on Mercury. Let's talk about the Sun's layers. As you might guess, the center of the Sun is super hot, with temperatures around 27,000,000 °F. This causes the atoms to slam into each other, forming larger particles and sparking thermonuclear fusion. As the hydrogen atoms join, they make helium, which is part of the nuclear fusion process. In a way, you can visualize the Sun as billions of atom bombs going off all the time.

As all of the Sun's energy and light bubble to the Sun's convection zone, many different forms of surface activity are triggered, such as flares and sunspots. After leaving the super-hot center of the Sun, the radioactive matter cools down a bit in the convection zone, where hot plasma bubbles. Ionized atoms make up the plasma. These atoms move up to the surface of the Sun. Now, when I say that the convection zone is "cooler," I mean that the temperature drops a bit below 3.5 million °F. Still, this is so hot that if you dropped diamonds into the plasma, they would melt and boil like soup!

Use the visors on your helmet to look closely at the surface of the Sun. Since it is so big, even the surface has a few layers. You might think that since they aren't as hot as the center of the Sun, we could at least get closer to the surface. I'd think again. The photosphere, chromosphere, and transition regions can get extremely hot.

Above the convection zone, the photosphere is the lowest layer of the Sun that we have directly observed. Around 250 miles thick, this layer of the Sun varies in temperature between 11,000 °F at the bottom and 6,700 °F at the top. The radiation from the photosphere makes the sunlight that we see on Earth.

Something strange starts to happen in the next layer, the chromosphere. Although we are moving away from the superhot core zone of the Sun, the chromosphere begins to heat up as the altitude climbs. The chromosphere starts around 6,700 °F at the bottom, but heats up to 14,000 °F the higher you go. This leads to a lot of surface activity on the Sun.

The last two layers of the Sun remain very hot. The transition zone is only 60 miles thick, but it skyrockets in temperatures up to 900,000 °F. Above it, the corona slowly tapers off into space. There is no upper limit, but as the heat dissipates out into space, the corona's temperatures start at the transition zone's temperature and often increase to a couple of million degrees Fahrenheit.

With all that heat, the Sun becomes a genuine danger. Between the chromosphere, the transition zone, and the outer corona, the Sun expels its heat. Here, we can observe the effects of the Sun's solar cycle and its sunspots, flares, and plasma streamers. Before checking out the Sun's surface, though, it's a good idea to double-check the Sun's cycle.

Did you know that the Sun has a special cycle? Every 11 years, the Sun undergoes a change in its magnetosphere. Thanks to centuries of people watching the Sun, we now understand a few things about the Sun's cycle. Since the Sun's cycle can affect Earth and technology, and astronauts in space, scientists are working to understand and recognize the clues that point to a new cycle starting. This way, we can prepare for high-impact Sun cycles.

Every 11 years, the Sun's activity ramps up. Scientists can predict the emergence of a new solar cycle by counting increasing sunspots. During this time, it might eject more heat and radiation than usual. When the sunspots appear around the poles of the Sun, it is a sure sign that something is changing inside the Sun. As new sunspots appear on or around the equator of the Sun, scientists recognize that the solar maximum has arrived.

After five years of buildup, the Sun shows extreme activity for a year, which scientists call the "solar maximum." During this period, the Sun might have more radiation eruptions and coronal ejections than usual. On Earth, we might be able to see more of the northern lights during this time. Thanks to the Earth's magnetic field, however, much of the Sun's ejections cannot hurt us severely. Meanwhile, the sunspots impact the Sun's magnetic field by expelling hot gas with an electrical charge. Do you know how a battery has a + and a - on each end? Sunspots have something similar. However, they aren't batteries for charging; they are ejections of energy that change the Sun's magnetic field. At the most active part of the solar maximum, the new magnetic ejections from the sunspots flip the magnetic poles so that the north goes to the south or vice versa. Pretty neat, huh?

Over time, usually a year, the Sun slowly calms down again. The new sunspots will outnumber the old sunspots, which tells scientists that the solar maximum is coming to an end. For the next five years, the Sun approaches the "solar minimum." Then, the cycle begins again. Even when the solar minimum begins, scientists can predict how strong the next solar maximum might be. In recent years, astronomers have started to observe the strength of the Sun's magnetic north and south poles. If the poles are weaker in charge, the next cycle will have a lower impact. However, if the poles have a strong magnetic charge, it's time to brace ourselves for some high activity during the solar maximum! Don't worry; we are safe for now. The space weather was part of the checklist before takeoff, or we would never have set out for Mercury.

You're free to peer out of the windows of the Dragon's Lance and watch the computer screen in front of you, and you'll catch glimpses of red flares erupting from the Sun's surface. Don't be alarmed. What you're feeling is the spacecraft reorienting itself to land on the surface of Mercury. See there, in the far distance, a beautiful view of the Caloris Basin. A 950-mile-wide crater edged with soaring mountains. It covers more area than the great state of Texas! Now our destination is coming into view—the Prokofiev crater. You see the faint, mysterious twinkle on the edge of the crater. Here you will stay for a few days to observe the small planet and the Sun.

I need all space travelers and crew members to please take their seats and fasten their seatbelts. Remain calm. The pilot and crew have erected the protective shields at the front of the Dragon's Lance. Suddenly exploding from the surface of the Sun, energy releases a stream of radiation into space along the path of magnetic field lines. The Dragon's Lance shudders, but the pilot and crew have already prepared for this by erecting protective shields at the front of the ship. All systems are still operational and online; we are clear to continue. What happened?

Ah—a solar flare! The captain saw an explosion on the surface of the sun. He prepared us for a small amount of turbulence. Often, the magnetic field lines near sunspots get tangled up, and as they reorganize or move apart, a surge of radiation will erupt into space. When these solar flares are powerful enough, they can interfere with Earth's radio waves.

Sometimes, a massive bubble of radiation bursts out of the Sun's atmosphere like a giant burp. These are called coronal mass ejections (CMEs). Since they are more significant than solar flares, they are mighty and can interrupt satellites and power grids on Earth. CMEs fling out enormous tons of plasma and radiation, sometimes up to a million miles per hour. However, these eruptions can take several hours to complete! The Sun isn't always erupting with solar flares and CMEs, but you are more likely to see them during the solar maximum.

Check it out! Something else is erupting off the Sun's surface! It is also very long and marvelous, looking like a fiery red loop. Beginning and ending in the photosphere, these prominences usually take a day to form and disappear, but some of them might remain for months. The large red loops are made of the Sun's plasma. Why or how prominences are formed is still a mystery to astronomers. Some scientists believe that prominences erupt when a magnetic field line becomes unstable, blowing the plasma for thousands of miles into space.

Between solar flares and prominences, you can see now why we can't just approach the Sun. Even if we are 1,000 miles away in the Sun's orbit, we might get slammed with plasma from a prominence or get hit by a wave of radiation from a solar flare. Yet, we want to look closer at the surface of the Sun, don't we? Let's get into the Observatory, where telescopes can help us view the Sun safely!

Once the Dragon's Lance docks, you can carefully move toward the Observatory with the rest. It takes a bit of time. Ok, kids, in this area, you can start testing out the new gravity of Mercury. Parents and crew, please read the safety guidelines before we exit the Dragon's Lance. Time to go! You can use a heavily protected external module and make your way to the now-open Observatory, where you will check out the Sun's surface weather. Race you there!

When looking through the solar telescope, you can see spicules on the Sun. These hair-like streams, spicules, wave off the surface all the time. Can you count the hairs on your arms? That can be pretty hard, right? Counting spicules is even more challenging.

To begin with, there are so many of them—up to 10 million at any single time. However, spicules don't last very long, so that makes counting them challenging as well. The longest a spicule will last is only around 10 minutes. They are fascinating to scientists. Some of them can extend very far off the Sun's surface, up to 6,000 miles long. The jets of plasma moving out can move as fast as 60 miles per second! Today, astronomers believe that spicules shoot out from the Sun because of different electrical charges interacting with the magnetic field.

See where I'm pointing? At that dark patch on the surface of the Sun. It looks like a freckle or mole; it's a sunspot! Sunspots look like tiny black spots on the Sun, but they can be pretty big. The telescope reports that the sunspots are cooler in temperature. Considering how hot the rest of the Sun is, the sunspots might seem colder, but they are still around 6,500 °F. Why are they cooler than the other parts of the Sun? Many astronomers believe that sunspots are seen around areas on the Sun where magnetic fields are very, very strong. These magnetic fields are so strong that they keep some of the Sun's heat away from the area, and the radiative zone's heat can't reach the surface. When magnetic fields clump together, their strength helps form cool sunspots, which can last months. Usually, this signals the close approach of the solar maximum.

Another beautiful and terrifying sight you observe through the telescope is "Sun rain." Do you think Sun rain is made from water . . . or from something else? If you are guessing something else, you're right. Sun rain is made from plasma that forms along magnetic lines above the Sun's surface and falls, following the magnetic field lines. During this time, the electrically charged electrons, protons, and ions of the rain are attracted to the magnetic field lines. Sometimes, Sun rain looks like it is coming out of nowhere, but it usually appears when a solar flare or coronal mass ejection happens. More plasma forms at the peak of the flare or CME and begins to fall back down to the Sun slowly. These coronal rain showers might last anywhere between 10 to 30 hours. Smaller ones last longer, but the most awe-inspiring Sun rain showers happen with sizable coronal mass ejections.

With Sun flares, prominences, spicules, sunspots, and Sun rain, the surface of the Sun is a pretty active place. As the Sun follows its 11-year cycle, it can have more or less activity, but in reality, the Sun is always moving, always changing, and always posing a danger to anyone or anything that comes close! The power of the Sun really does help us appreciate the defenses and protection of the Earth's magnetic fields.

That's right. The Sun isn't the only member of the solar system to have a magnetic field. Some planets, like the Earth, have strong magnetic fields; others have barely any. On the other hand, the gas giants of our solar system have strong magnetic fields.

Thanks to the Earth's magnetic field, the impact of the Sun's solar flares and coronal mass ejections are lessened. Without the Earth's defensive magnetic field and thick atmosphere layers, our computers and phones would lose all of their information and no longer work! We wouldn't be able to grow plants or hold water either. Sadly, the Earth would probably look more like Venus or Mars: too hot and unable to sustain life. The interaction between the Sun and our Earth's magnetic field is largely invisible, but you can see the impact of the Sun on our Earth's magnetic field on some special occasions.

Have you ever been to Alaska, Iceland, Norway, or the northern regions of Canada? If you travel up to those areas in 2024, you will be able to see a fantastic display in the sky: the aurora borealis. One of the best ways to look up and be reminded of Earth's excellent defenses is the aurora borealis, which usually increases as the Sun's solar maximum increases in activity. Although the aurora borealis can be seen in the north, the aurora australis, closer to the magnetic South Pol,e reveals the same phenomenon.

After a solar flare, the solar wind might cause some of the Sun's particles to blow across the solar system. Due to the weaker magnetic fields found at Earth's magnetic poles, the Sun's particles sometimes enter the Earth's thermosphere and collide with the Earth's atmospheric particles. The aurora borealis, also known as the "northern lights," never appear any lower than the thermosphere. Still, around midnight, we can see them quite clearly!

The northern lights are beautiful. They look like streams of light or rippling curtains of light in the night sky. The primary color of an aurora borealis is usually pale yellow-green, but sometimes it might be red, blue, or purple. The different colors depend on which gas particles are colliding together. When the solar flare is extreme, the northern (or southern) lights will change color constantly. It will look like a gentle ripple or even a dance. This causes a beautiful, silent, and eerie light show that has inspired the imagination of many observers for thousands of years. That's the power of our Sun!

However, our Sun isn't the only powerful Star in the sky. There are billions of stars in the Milky Way. The closest neighboring solar system to our Sun holds many mysteries: the Alpha Centauri system. Alpha Centauri has two stars that orbit each other: Rigil Kentaurus and Toliman. They are a bit older than our Sun. Rigil is bigger and brighter than our Sun, and Toliman is a bit smaller and more orange. However, there is something else strange in this system: a third orbiting Sun!

Smaller and redder than our Sun, Proxima Centauri is classed as a red dwarf. The naked eye cannot see this small star, although you can see Rigil Kentaurus and Toliman from continents in the Southern Hemisphere

(Australia and Africa, for example).

Orbiting Rigil Kentaurus and Toliman, Proxima Centauri takes 550,000 years to circle its two neighboring suns. Due to its wide orbit, Proxima Centauri is the closest star to us. Traveling at the speed of light would take us just over four years to get to Proxima Centauri, that is 24 trillion miles away! As you can imagine, getting to Proxima

Centauri would take forever. With the methods of travel that we have today, it would take over 6,000 years to get to Proxima Centauri. For now, we will have to be content with just observing Alpha Centauri through telescopes.

Rigil Kentaurus, Toliman, and Proxima Centauri might be too far off to explore. However, we can still use new technology, satellites, probes, and new astronomical discoveries to observe bright stars in our Milky Way galaxy. Humans have used stars as the foundations for constellations for thousands of years, connecting the dots to make pictures and tell stories.

Today, the leading stars of those stories, like Polaris, aka the North Star, are observed by scientists. Polaris, Sirius, Betelgeuse, and Rigel are all far beyond our reach. Still, through observations and analyses, we can better hone our understanding of math and the orbits and different natures of stars. Meanwhile, on our own doorstep, we can appreciate the power and beauty of our own Sun. Thanks to its light and energy, our Earth can sustain life.

Now that we've visited the Sun, it's time to get up close and personal with Mercury, Venus, Earth, and Mars!

CHAPTER 6

TERRESTRIAL TERRAINS: THE INNER PLANETS

Well, let's look out some of these windows from which we can view Mercury. Look out of the window to your right. You can see that Mercury looks a lot like our moon, with gray dust and craters. That's because Mercury's outer shell has similar characteristics to the Earth and the Moon. However, its inner core is metallic with little to no molten qualities. Both the core and the outer shell of Mercury are thinner, so overall, the planet is not much bigger than the Moon.

Mercury rotates perfectly up and down on its axis, so it does not experience seasons as other planets do. It does experience extreme ranges in temperature despite its nearness to the Sun. What's happening? It's all to do with Mercury's rotation. Although it moves around the Sun very quickly, the orbit of Mercury is moderate. With such slow rotation, you can imagine how long Mercury's days can get! Amazingly, 59 Earth days make up one day on Mercury, which means that one side is scorching for long periods and the other side is frigid. With no atmosphere and weather, standing on Mercury in the daytime can get as hot as 800 °F. At night, it drops to -290 °F. That's pretty extreme, right?

If we get into special suits, we might be able to walk on the dark and colder side of Mercury. Once you walk on the planet, you immediately notice something different. Each step feels lighter. That's because Mercury's gravity is lower than Earth's—at only 38%! You will feel less than half your weight, so if you are 100 pounds on Earth, you'll only weigh 38 pounds on Mercury!

We will hike for half an hour to reach the crater's edge, where a mind-boggling surprise awaits. Make sure you don't go too far!

Without our communication link, we won't hear each other on Mercury. Without much of an atmosphere, sound can't move, so talking on Mercury is near impossible.

Looking around, what do you see? Not much, right? That's because, without an atmosphere, water, and adequate regular heat, Mercury cannot naturally support life as we know it. For now, Mercury is empty. Let's have a look at some primary data, so we are more familiar with this small, superhot planet.

- **Planet Type: Terrestrial Planet**
- **Distance From Earth: 0.61 AU**
- **Distance From the Sun: 0.4 AU**
- **Radius: 1,516 miles (⅓ the size of Earth)**
- **Rotation (Solar Day): 59 Earth days (6.7 miles/hour)**
- **Orbit Around the Sun (a Year): 88 Earth days**
- **Moons: 0**

It's incredible how many craters exist on Mercury; scientists believe the planet was struck by a meteor around the time the solar system was created. Perhaps it was an impact like this that took off Mercury's outer shell. To this day, Mercury still gets bombarded by meteors. Without atmospheric protection and moving so close to the Sun, the tiny planet is a prime target. Perhaps there will be a new basin to observe one day!

Now we've arrived at the edge of the crater. Here, in the deep shadows, sunlight never reaches, so the ice water deposits glint in the dim light. Ice water on such a hot planet! It's hard to imagine, but given Mercury's slow rotation, it's understandable how this cold water has been kept safe.

Since ancient times, Mercury has been an old friend in the night sky. With binoculars, you can spot Mercury in the night sky just after the Sun sets or before the Sun rises. It will appear close to Jupiter. You can find it in the lower reaches of the horizon an hour after the Sun sets in the western sky. If you look for it in the morning, you can find it in the eastern sky close to the horizon as well. From May 3rd to May 24th, Mercury will be bright and easy to spot in the evening sky. Between October 18th and November 1st, you can spot it more readily in the morning sky.

Mercury was an exciting planet to see, wasn't it? I bet that Venus is even more exciting. Let's go visit Venus and find out whether it is as beautiful as the goddess it was named after!

As we fly close to Venus, we notice right away how its yellow-white glimmer looks enchanted. It certainly looks like it might be a magical planet to land on. After all, it's around the same size and shape as our Earth. If we cut Venus open and compared it to the center of the Earth, it would look similar as well, with an iron core, a thick molten rock mantle, and a thinner rocky outer shell. In fact, Venus might be experiencing "plate tectonics" like Earth does, which may also be reshaping Venus. That's only the beginning of what I want to tell you. Okay, we are landing at the Venus Observatory, located close to the top of Maxwell Montes. As we gaze over this stunning sight, let's also check out some information on Venus.

- **Planet Type: Terrestrial Planet**
- **Distance From Earth: 0.28 AU**
- **Distance From the Sun: 0.7 AU**
- **Radius: 3,760 miles (just a bit smaller than Earth)**

- **Rotation (Solar Day): 243 Earth days (4 miles/hour)**
- **Orbit Around the Sun (a Year): 225 Earth days**
- **Moons: 0**

Wait. Something doesn't look right. Have you noticed those clouds as we began to descend? What does the ship's instrumentation say? Wow! The temperatures on Venus are beginning to skyrocket even higher than on Mercury. The instrument readings are at 900 °F! We will undeniably need to wear our protective spacesuits. I want to tell you why the temperatures are so hot, and the clouds are visually yellow. Scientists have realized that Venus's air is very poisonous since it is mostly carbon dioxide. On Venus, its carbon dioxide atmosphere, its closeness to the Sun, and its weak magnetic field have turned it into a piping hot, poisonous planet.

Visually, the Sun comes into question. Do you see that kind of bright smear of light over there to your right? That's the Sun! You can't see its shape because the atmosphere of Venus is so thick. Also, you might have noticed that thanks to Venus's retrograde rotation, the Sun sets in the east and rises in the west. Day and night feel like they are lasting forever on this planet as well. This is because Venus rotates super slowly, taking around 243 Earth days to pass before a new day starts here. That's more than half a year on Earth! Please note that with Venus's thick atmosphere of sulfuric acid clouds, you might not notice sunrise and sunset as simply as you can on Earth. All of this tells us that there is a reason why we haven't found life on Venus. For an organism to survive on Venus, it wouldn't be anything we would recognize on Earth.

We are now descending to make landfall underneath the cloud cover of Venus. Everybody will be shielded by the durable, thick, and insulating materials of the Dragon's Lance. It's not just protecting us from the toxic environment of Venus, but it is also highly pressurized to protect you from the weight of the atmosphere. Since Venus is close in size to Earth, walking on Venus actually wouldn't be too difficult. If you were 100 pounds on Earth, you would weigh in at 90 pounds here on Venus. The difference isn't that big. However, the atmosphere is so thick that it would feel like you are swimming two-thirds of a mile deep in the ocean. The weight of the atmosphere on your arm as you waved hello to your friend would remind you that you are no longer on Earth . . . if you had time to wave hello to your friend. More than likely, you would be crushed and burned up within seconds. Not today though! Thankfully, our special EMU suits will help us as we go for a short walk on Maxwell Montes. This is where we can enjoy the Venetian heights.

The carbon dioxide skies on Venus form cool-looking clouds in the yellow skies. However, these deadly Venusian clouds are made of sulfuric acid, a chemical compound that can burn you. Whether it is a gas or a liquid, sulfuric acid is not suitable for your body. On Earth, sulfuric acid is often used for manufacturing, but it can burn you if it touches your skin. Breathing in sulfuric acid burns your lungs as well. As you can imagine, this makes Venus's weather even deadlier. If rain were ever to fall on Venus, it would be sulfuric acid rain!

One of the mysteries of Venus is how active its molten layers are. Although scientists have not yet sighted recognizable volcanic activity on Venus, many clues point to a long history of volcanic activity on Venus. Unlike Mercury, Venus has diverse terrain, with one mountain higher than Mount Everest on Earth—that's Maxwell Montes. We will walk across a small section of the side of this massive volcano mountain. In the distance, you might be able to glimpse other craters and canyons as well as Venus's signature "pancake" domes and "tick" domes. These shapes were created by lava exploding out and then moving slowly down and across the terrain. However, as far as scientists can tell, these lava floods didn't happen recently. There's more to see. Make your way back to the Dragon's Lance, where we can look at more information in the Observatory module.

This is where you can visit the Venus Exhibition. These displays are of the early video and photographs taken by the more successful Russian probes that landed on Venus. Venera 13, after about two hours, was finally crushed and destroyed by the atmosphere on Venus. However, Venera 13 managed to analyze some soil and take photographs, sending them to Earth before it went offline. Thanks to the combined efforts of space agencies around the world, we are learning more and more about Venus.

Since humans first began to notice the stars in the sky, the bright glimmer of Venus has always stood out. Often called "the morning star," the Romans named Venus after the goddess of beauty. The Romans didn't know that although Venus looked beautiful from a distance, her bright clouds were poisonous.

Today, you can still find Venus in the eastern sky in January, about an hour before the Sun rises. Until late May, Venus is too close to the Sun to be seen, but you can find it in the western sky around sunset after late May until the end of December. With binoculars, you will be able to notice more detail on December 5, which is when

Venus shines the brightest. It will also start changing shape, looking more like a crescent moon.

After spending the day on Venus, I'm guessing you're getting excited about seeing Mars next. First, we'll return to Earth's orbit, hang out at the International Space Station, and visit our old stomping grounds: the Moon.

As we get closer to Earth, you might notice a long, thin object floating in orbit, with eight thin panels sticking out on either end. In the middle, tubes and cargo bays open up into various modules. This is the International Space Station, a decades-long project that many countries have teamed up to work on. Watch as we dock up. You will notice that there are different docking compartments for various spacecraft to attach to the space station. We can safely pass from the Dragon's Lance into the International Space Station (ISS) through these airlocks.

Some parts of the space station are older than others. The first piece of the ISS was launched into space by the Russians. In 1998, it was smaller, but over time, more elements were added by America, Russia, Japan, and Europe. Many countries added more modules for lab work and experiments. Some modules included living areas for the astronauts staying in space for long periods.

The ISS has a lot of expensive and essential equipment, so we must be careful of our surroundings as we float around. If you look out of that window, you can see even more technological devices outside, like the robot arms used to build other parts of the space station or the solar array panels. On both ends of the ISS, solar arrays stick out, collecting the sun's rays and converting them into electricity.

As you can see from the narrow corridors and the focus on experiments and work, the ISS isn't a place for tourists. Here, experiments and deep space observations are made as astronauts help prepare for more extended explorations into space. Let's stay out of their way by this window and look down onto Earth. Can you imagine living here day after day?

Since the ISS isn't on a planet, there isn't any real force of gravity inside. Living in zero gravity (zero-G) can have its ups and downs. Although some of the stuff you can do with food is cool, and floating seems like less of a hassle than walking, simple routines like washing your hair can be complicated.

As a result, scientists and engineers have had to invent new ways of washing your hair, going to the bathroom, and eating your meals. For example, there are no refrigerators in space, so you can't store food for long durations. It's kind of like camping!

Here, check out this brownie. It tastes just as delicious in space, doesn't it? Some foods, like pasta, can have water added to them and then heated in an oven. Fruit and bakery snacks, however, can be eaten just the same way. After you finish eating your brownie, we need to dispose of the packaging carefully. Just like on Earth, we need to be careful about what we do with trash.

After our healthy meal and delicious snack, we can hang out by the windows and watch the Earth turn. There are movies to watch and games to play, and we can even chat with friends and family back on Earth! However, you may notice that the astronauts have to do quite a bit of work. If they aren't working, they take time off to rest and relax. Since living in zero Gs all the time isn't healthy for your body, the astronauts on the ISS have to exercise at least two hours a day as well! Don't worry, though: We will only stay here for one night, so we will be fine!

Yep! You heard that right. We will dock with the ISS for one night, but we'll be bunking in our own Dragon's Lance. In this weightless world, we can't lie down on beds. We would just float off! You will have to sleep in a one-person crew cabin in a specialty sleeping bag. Tomorrow, we will be visiting the Moon Observatory. Goodnight, now you can enjoy drifting off to sleep in zero-G while watching the Earth pass by.

Many people can view the space station on different days from Earth, usually around sunrise or sunset. Depending on where you are standing and how high up you are, you can find the ISS above the horizon for a few minutes. Due to its close orbit of the Earth and its speed, the ISS will travel across the sky and disappear after a short period of time. Usually, it is cloaked by the Earth's shadow or the horizon.

Although you can see its glow with your naked eye, a telescope could give you a bit more detail. If the setting sun rays are strong enough, the light will bounce off the ISS and give you a blurry look at Earth's largest man-made satellite. If you want to find the ISS station while on Earth, check out NASA's "Spot The Station" website to find out where you might be able to see the ISS station from your neighborhood!

After visiting Mercury and Venus, Earth looks really cozy, doesn't it? It's time to head to the Moon, though. As we watch the Earth shrink into the distance, the white swirls of clouds and the blue, green, and soft browns of the Earth's surface feel inviting. Down on planet Earth, plants, land, ocean creatures, and humans live together and thrive. It makes you wonder what Earth has that makes it the perfect place to sustain life.

When analyzing the structure of Earth, we notice that it looks a lot like Venus, with a solid iron core covered with molten rock and then a thin outer shell of rock and soil. However, unlike Venus, Earth isn't too close to the Sun: It rotates on a slight tilt that creates seasons, and its magnetic field protects the planet from the worst heat of the Sun's solar flares and coronal mass ejections. Since the Earth's orbit isn't perfectly circular and its rotation isn't too fast or too slow, the equator generally stays warm, and the areas closer to the North and South poles experience seasonal changes. This "Goldilocks Zone" allows the Earth to sustain many life forms.

Since the Earth's rotation and orbit are stable and perfectly positioned in relation to the Sun, the day-night cycle isn't too long or short. Also, the day and night cycles, for the most part, are balanced to allow for proper amounts of sunlight. Closer to the poles, the day-night rhythm on Earth changes. For many months of the year, people living far in the north, such as in northern Alaska or the Arctic, will experience sunlight all day.

Thanks to the harmony of these conditions, Earth has frozen wastelands and boiling deserts, but even though we find these areas difficult to live in, these regions have plants and animals that have learned to adapt. The changeable weather on Earth can get dangerous, spawning tornadoes, lightning storms, and hurricanes. Overall, the weather won't poison us like Venus's deadly clouds, nor will it freeze us or boil us to death like Mars. This is all thanks to the Earth's magnetic field lines and multiple layers of atmosphere, which protect it from space debris and the Sun's radiation. Indeed, our home planet is incredible! Here's a quick refresher on basic Earth information:

- **Planet Type: Terrestrial Planet**
- **Distance From the Sun: 1 AU**
- **Radius: 3,959 miles**
- **Rotation (Solar Day): 24 Earth hours (978 miles/hour)**
- **Orbit Around the Sun (a Year): 365 Earth days**
- **Moons: 1 (the Moon)**

Watching the Moonrise is easy on Earth, so humans have always noticed it, and for thousands of years, ancient civilizations have tracked the phases and movement of the Moon. The word "month" comes from its name as an indication of a way to tell time. For some cultures, the lunar calendar is almost as important, if not more so, than the solar calendar! It's no wonder, then, that humans began to look more closely at the Moon as soon as early telescopes were invented. Today, a powerful pair of binoculars can help you appreciate the surface of the Moon.

We won't need binoculars today because we will be landing on the Moon!

As we draw closer, you will notice dark patches on the Moon. Scientists believe these patches were lava, once upon a time, that have since become solid. On Earth, they look like mysterious dark areas, which are often called "lunar seas." However, there is no water here on the moon. On the Mare Tranquillitatis (Sea of Tranquility), Neil

Armstrong took humankind's first step onto the moon. There are many other lunar seas: Mare Imbrium (Sea of Showers), Mare Serenitatis (Sea of Serenity), and Mare Nectaris (Sea of Nectar).

After our spacecraft lands, you can get into your EMU suit to prepare for our moonwalk! Once out on the surface of the Moon in our space suits, you will notice something right away. Walking, running, and jumping on the Moon is super easy. That's because the Moon is smaller and has less mass than Earth. With little to no gravity, you will feel very light on the Moon. If you weigh 100 pounds on Earth, you will only weigh 16 ½ pounds on the Moon! That makes moonwalking positively fun and a must-have experience.

However, beyond the Dragon's Lance and the fantastic view of Earth from the Moon, there isn't much to see here. Without an atmosphere, there is no oxygen on the Moon. There is no sound. No wind. No plants or animals. In fact, it is so still on the Moon that our footprints on the ground could stay for millions of years.

Another mesmerizing sight we can see on Earth is eclipses. Solar eclipses happen when the Moon gets between the Earth and the Sun. Depending on where you are and what date it is, the solar eclipse might be total or partial. Since eclipses rely on patterns and cycles that astronomers can mathematically calculate, we can easily predict eclipses! Have you seen an eclipse yet? Perhaps one day you will! If you do plan on watching an eclipse, though, be sure to put on protective eyewear. Even when covered by the Moon, the Sun can damage your eyes.

Ever since ancient civilizations noticed the patterns of solar eclipses, humans have followed the path of the Sun and the Moon.

Many cultures saw the dance of the Sun and the Moon as exceptional or as a warning. Today, we enjoy eclipses as a reminder of nature's conscious cycles.

When we spend time in nature, we can better appreciate the beauty of the universe, not just our planet. When we go out on a clear night and look up at the stars in the countryside, we will see the night sky more clearly than we would in the city. The entire sky is covered with brilliant speckles. However, you might notice some stars look brighter than others. Sometimes, they appear to be in groups. You might be tempted to connect the dots!

You aren't the only one who wants to make shapes with the stars. As long as human civilization has been around, individuals have always made shapes out of the brighter stars, giving the groups of stars names and telling stories about what they mean. Today, scientists call these groupings "constellations." There are around 88 officially recognized constellations. However, depending on where you live on Earth and what time of year it is, your view of the constellations will be limited. Some famous constellations include the Big Dipper, Orion, the Big Bear, and the Northern Cross. Learning all of the constellations and discovering them in the night sky can be a lot of fun!

We have to keep on schedule, so it's time to head for the famous red planet, Mars. It's a bit of a trip, but as we approach Mars, you get a clearer view of the reddish-brown planet. It isn't as big as Earth, but it does look similar in size. However, there are no swirls of clouds, and there are no signs of animal or plant life as we land. Our spacecraft is now approaching the Martian surface, which looks rocky and dusty. This is Mars, the first planet humans will settle on beyond Earth. For a long time, various probes and rovers slowly and carefully explored Mars.

For many decades, scientists have analyzed Mars and recognized the many similarities between Earth and Mars. The rotation of Mars is similar to Earth, and each Martian day is around 24.5 hours long. That feels pretty familiar. Like Earth, Mars's core is made of hard rock, more than likely iron, sulfur, and nickel. Around the core, there is a thick rocky mantle. On top of the mantle, iron, aluminum, and other materials form the crust. On Earth, our crust consists of rock, soil, or sand. Mars, however, has mostly reddish earth because of iron in the soil that rusts thanks to the thin atmosphere. Other interesting landscape features on Mars include volcanoes, impact craters, and canyons.

Mars might feel familiar, but before we go out for a walk on Mars's giant volcano, we need to suit up. That's because the atmosphere is made up mostly of carbon dioxide, argon, and nitrogen, which we can't breathe. With oxygen tanks and securely fastened helmets, we will be ready to walk on Mars. Remember to dress quite warmly! Since Mars is further away from the Sun and its atmosphere is thinner, its temperatures tend to be lower. Your feet will feel warmer than your head since heat escapes from the Martian surface more quickly. Thanks to its orbit and the tilt to its axis, Mars does experience four seasons, but they last longer than Earth's because the planet's solar year is also longer. Furthermore, the only real difference between seasons is the temperature changes. At its warmest, Mars is only 70 °F, but it can drop as low as -225 °F. Brrr! Unfortunately, since Mars has no magnetic field, it doesn't have much protection from the Sun. It only keeps cool because of its distant orbit.

Also, Mars has a lower mass than Earth and therefore has less gravity. This affects how much you might weigh on this planet. For example, if you are 100 pounds on Earth, when you stand on a weight scale, you will only be around 38 pounds! Walking on Mars feels lighter and more effortless. You probably have noticed the difference as well!

Today, the search for life on Mars continues. Although scientists have found evidence of frozen water on Mars, there has been no discovery of bacteria or other familiar life forms. Although Mars might be safer to build a human settlement on, it is far from being ready to support human life—or any kind of life—without the aid of technology and science.

- **Planet Type: Terrestrial Planet**
- **Distance From Earth: 0.52 AU**
- **Distance From the Sun: 1.5 AU**
- **Radius: 2,106 miles (½ the size of Earth)**
- **Rotation (Solar Day): 24½ Earth hours (538 miles/hour)**
- **Orbit Around the Sun (a Year): 1.88 Earth years**
- **Moons: 2 (Phobos and Deimos)**

I think we need to have a quick flight around Mars. There is so much to see! Have you ever been to the Grand Canyon in America? It's a supermassive canyon with a large trench, but Mars has an even bigger one. Called Valles Marineris, this canyon is so long it could stretch from California to New York! It's over 3,000 miles long.

On top of that, it is broader and deeper than our Earthly Grand Canyon. Around 200 miles wide and 4.3 miles in its deepest area, Valles Marineris is one of the great wonders of our solar system. Let's fly our spacecraft along the canyon and admire the steep ridges. It is mind-blowing how substantial this space is! And it's outright empty. I wonder if, one day, Martian farms and cities will fill this blank trench.

Oh, look! Now we are headed to the next incredible sight on Mars: the solar system's biggest volcano. Its name is Olympus Mons, and it is around the same size as the entire state of Arizona. That's massive!

Look closely on your right as the spacecraft swings around. We are going to land at that observatory near the top. Don't worry. As a dead shield volcano, Olympus Mons hasn't been active in a long time. You may have noticed that it looks like a large, flat bump on the surface of Mars. This is because when it erupts, the lava just slowly slides down the sides of the volcano instead of erupting violently. Still, Olympus Mons is a very quiet volcano. Today, we are still studying whether there are hidden ice deposits beneath the surface.

Olympus Mons holds many secrets!

Without an atmosphere, I want to tell you about the Martian weather. The red planet does have a few dust storms, where winds will begin to whip up in circles like a dust devil or tornado. However, because the atmosphere is so thin, the breezes on Mars are pretty gentle, moving only between 10 to 20 miles an hour. The highest recorded wind speed was around 70 miles an hour. This means that, for the most part, the wind and weather of Mars are easily manageable.

Once every five years or so, Mars might have a planet-wide dust storm.

Without greater gravity, the dust may hang in the air for many days in a row.

During this time, keeping technology completely free of dust is impossible. With technological advancement, we can create ways to protect machinery on

Mars from these dust storms. Comparing the hurricanes and tornadoes on Earth, Martian storms are not as big of a problem for human settlement.

Today, the weather is clear, so we can soar gently over the mountain slopes of Olympus Mons.

As we fly by, look out of the Observatory module, and be sure to look up and to the left. You will be able to see one of the two captured asteroids, Phobos and Deimos, float by. Check them out with this telescope. You can see that they are not really moons but asteroids shaped like potatoes. Landing on these moons isn't easy because they are so small, so we will just watch them whirl by instead. Named after the horses that pulled the chariot of the god Mars, these two minor asteroids orbit Mars rather quickly. Phobos circles Mars three times a day, while Deimos completes an orbit every 30 hours. Due to the speed of Phobos, scientists believe it will one day crash into Mars or disintegrate into a dust ring. Mars still has a lot to show us.

It has always been visible to the naked eye thanks to its reddish glow. The ancient Egyptians and the Romans noticed this color difference right away.

Linking the planet to the god of war, the Romans named it "Mars." Today,

Mars is still apparent, and with the help of a telescope, more details can be seen.

You can find it in the evening between January and August and in the morning between November through December. Since Mars's orbit isn't always close to the Earth's, it often seems to disappear from the sky between August and

November. However, you should find it easily by the moon in May and by Venus in July.

Standing on the slopes of Olympus Mons, looking down the massive slopes over the dust devils whirling in the distance, Mars feels both alien and familiar. A home away from home, Mars is our first and most logical planet to become a multi-planetary species. However, there are still mysterious planets and moons to explore in the outer solar system!

CHAPTER 7

GASEOUS GIANTS: THE OUTER PLANETS

Touring the outer system and seeing the gaseous giants' icy beauty is magical and more dangerous. Since these planets are very far away from Earth, we will need to be especially careful as we travel. Our journey is only halfway made! Are you ready to explore the gas giants?

On our way to Jupiter, we will pass by the asteroid belt. Keep an eye out!

We can pause here for a moment in the asteroid belt, and we can float over the Observatory module and look around. Whether you look left, right, or up through windows, you will notice that there are rocks of all shapes and sizes flying about. Do you know what these are? Asteroids! They probably came from planets that broke up millions of years ago. This is why most asteroids are made of iron and nickel. As we approach one of the asteroids, you will notice that it looks pretty plain—a basic gray or brown color.

Nowadays, asteroids speed along at around 15 miles per second. However, not all asteroids move at that speed. Some move slower than others, and many of them orbit at a great distance from other asteroids. The space between asteroids is so vast that if you stand on an asteroid, you won't see the other asteroids orbiting with you.

Landing on an asteroid is possible, but we will observe one of the larger ones up close. We can't land on all asteroids. Some of them are the size of pebbles, while others are huge. One of the largest asteroids, Ceres, is categorized as a dwarf planet.

Now we are smoothly pulling away from the asteroid belt. Leaving the familiar behind us, we move into the mysterious world of the outer solar system. Here, the Sun and the Earth are far, far away, and an eternal winter reigns. Still, even here, there are peculiar sights, puzzling mysteries, and new worlds to discover.

Now coming into view—Jupiter! This massive gas giant dominates our ship's touch screens, and outside our windows, it fills up the sky. All we can see are its marbled swirls, looking like whipped ice cream. With the brilliant colors of red, orange, white, and blue, Jupiter's clouds churn together in beautiful bands. However, as we draw closer, you'll see that we are looking at clouds moving along at incredible speeds. There is no solid surface to walk on.

As we sink into the clouds of Jupiter, notice that the instruments on the Dragon's Lance are warning us that the winds are moving as fast as 335 miles an hour. The Great Red Spot on Jupiter is another storm that has been raging for over 300 years. Although there are no straightforward theories as to why these storms form, scientists still work to analyze the data we do have about Jupiter.

Hanging above Jupiter in orbit, we can see many of its moons, with Io, Ganymede, Callisto, and Europa being the largest. After we finish flying through the upper reaches of Jupiter's clouds, we'll visit Jupiter's moons. For now, let's admire the lightning storms and the heavy vapors of Jupiter's atmosphere. If we went outside right now, we'd not just be battered by the wind, but we would need our suits to breathe. These clouds are made of icy ammonia. That's like inhaling fertilizer or cleaning spray—not great for your lungs. Other lightweight materials, such as hydrogen and helium, also make up Jupiter's atmosphere. Although we have not yet been able to find Jupiter's center, it is theorized that Jupiter might have a massive hydrogen ocean or plasma in its center, like the Sun. The pressure in the center may be so great that Jupiter's superfast rotation and electric current within the pressured hydrogen forms its massive magnetic field.

Since Jupiter spins around so fast, its day cycle is only about 10 Earth hours long. That's a short day! However, its orbit around the Sun is so sweeping; one solar year on Jupiter is just under 12 Earth years! Because it barely has a tilt on its axis, it has a strong magnetic field, and its orbit is so far from the Sun, Jupiter doesn't experience seasonal weather changes. Instead, it appears to have a stable surface temperature of -160 °F. It might be warmer at its core, but for now, scientists can only theorize what is literally happening beneath Jupiter's thick gaseous clouds.

Strangely enough, even though Jupiter doesn't have land to walk on, the pressures below the clouds are too much. With Jupiter's size and mass, it's no surprise that gravity is staggering here as well. If you weighed 100 pounds on Earth, you'd weigh in at 250 pounds! Thank goodness we can't walk on

Jupiter!

Thanks to gravity, the lack of solid land, and the high wind speeds, life on Jupiter is all but impossible. Even setting up a space station in orbit is difficult because Jupiter's magnetic field traps electrically charged particles and speeds them up, creating high radiation. Overall, Jupiter isn't a great place to live, but it sure is a breathtaking planet to explore! Check out some of these fun facts about Jupiter!

- **Planet Type: Gas Giant**
- **Distance From Earth: 4.2 AU**

- **Distance From the Sun: 5.2 AU**
- **Radius: 43,441 miles (11 times the size of Earth)**
- **Rotation (Solar Day): almost 10 Earth hours (28,324 miles/hour)**
- **Orbit Around the Sun (a Year): 11.86 Earth years**
- **Moons: 79 (largest: Io, Europa, Ganymede, and Callisto)**

Thor. Amirani. Chaac. Prometheus: These are all of Io's striking volcanoes. As we land on Io, we need to be very careful! Before you leave the Dragon's Lance, you will need to check with the Orion Tour crew. On Io, you never know when a lava flow is going to swamp you! Thanks to the orbits of Europa and Ganymede, Io's orbit is very irregular, which causes tidal forces to push and pull on its surface. On Earth, we experience tides that push and pull water up to 60 feet. On Io, the solid ground bulges in and out as much as 330 feet!

With all of this movement, it's no wonder that Io is the solar system's volcano hotspot. If you love volcanoes, you are going to love Io! Come over here and check out this lake of lava. It was newly created a few weeks ago. Over time, this will harden, but a new volcano or lava spill will happen again in a few orbits. Now you see why staying on Io is not a reasonable option. Thanks to the toxic atmosphere and volcanic activity, we will swiftly probe the unstable planet and continue our quest.

Our next stop will be Ganymede, Jupiter's largest moon. It is the largest moon in the whole solar system. Ganymede has its own magnetic field, which results in some small amounts of glowing gas at its north and south poles. Just like the aurora borealis on Earth, these atmospheric changes usually happen when foreign particles bounce in and out of Ganymede's thin atmosphere or when Jupiter's magnetic field lines change. However, unlike Earth, Ganymede's auroras are very faint.

Although Ganymede is a solid ice and rock planet, scientists believe they have found evidence pointing to a hidden saltwater ocean underground. This ocean might hold more water than what we have today on Earth! As a result, Ganymede is considered a great alternative base for human settlement after Mars. However, Ganymede is so far away from the Sun that this moon is super icy cold! The ocean of water might be frozen over. After dinner, we can go for a short walk in our EMU suits over Ganymede's wintry, rocky landscape. Look up and admire Jupiter's thin rings and swirling clouds!

The show's real star is Europa, where we will stay the night and enjoy a fantastic view of Jupiter. Just a bit smaller than our Moon, Europa shines brightly thanks to its glittering surface of frosted ice water. Scientists believe Europa might have ice slush or flowing water underneath miles of ice, which could hold very primitive life forms. What would they look like? Nobody knows. For now, there are many worries about how to analyze Europa's waters without contaminating its biome. In recent years, telescopes have caught plumes of matter spraying off Europa into space, so many scientists consider these geysers a new way to analyze Europa's water.

As our spacecraft draws closer to Europa, you will notice the lines and cracks all over Europa's surface. This shows that Europa's icy shell is also affected by tidal forces with Jupiter. Don't worry, though! As we settle down on the thick ice surface, the Dragon's Lance won't sink through the ice. It's at least 10 miles thick. With so much ice, you probably think Europa is the solar system's ice rink. It kind of is! As we enjoy the vistas of Europa, be sure to take in the memorable sight of Jupiter as well. It always stands out.

Jupiter has always been noticed, in history, by other civilizations. Named after Jupiter or Jove, the king of the Roman Gods (equivalent to Grecian Zeus), Jupiter is the greatest of the solar system's planets. Compared to the

Sun, it's noticeably small, but Jupiter dwarfs Earth. As the bulkiest planet, Jupiter looks like a silver-white star from Earth. From the end of August to the first week of January, Jupiter shows up in the night sky. From February through August, you can also find Jupiter low on the horizon by Mercury before sunrise. With the help of binoculars, you will be able to see Jupiter easily, and with a good telescope, you might even be able to catch a glimpse of the different bands of colors!

Jupiter and its moons were pretty fun. You may wonder if Saturn or the other planets will be just as spectacular. Wait until you get to Enceladus!

Saturn is pretty imposing as the second-largest planet in our solar system, but what makes it so dazzling are its wide rings and mysterious moons. Covered with clouds like Jupiter, it has faint bands of yellow, brown, and gray. With clouds in the upper atmosphere swirling around at 1,600 feet per second, Saturn's wind speeds make hurricanes on Earth look like slowpokes.

The pressure on this planet is also very high, so toward the center, it is possible that we would find the gas squeezed into a liquid.

That means we would probably be squished flat if we flew into the center of Saturn. It doesn't have crazy amounts of mass. The average density of matter on Saturn is less than water. If you could find a big enough bathtub, fill it with water, and put Saturn inside, Saturn would float! With this in mind, it's understandable why, despite its size, gravity on Saturn would be closer to Earth, making you only a bit heavier. If you could walk on Saturn and you were 100 pounds, you'd weigh around 107 pounds! That's not bad, right?

However, living on Saturn is almost impossible. To start with, Saturn doesn't have much of a landmass. Scientists theorize the center may have a small dense core of iron and nickel, but covered with liquid hydrogen. The fluid hydrogen points to another problem: The atmosphere is toxic. Made of hydrogen and helium, you wouldn't last long breathing in the "air" even if you could survive the winds somehow. On top of that, Saturn's distant orbit around the Sun promises super chilly weather.

Interestingly, Saturn tilts on its axis like Earth, so it probably experiences seasonal changes, but I'm sure that Saturn's seasons look nothing like Earth's. Saturn is so far away from the Sun; it takes sunlight 80 minutes to get from the Sun to Saturn. This means that Saturn and its moons are sure to be crazy cold! As we fly close to the clouds, we realize that the heat from friction and pressure at the core keeps Saturn going, but beyond that, it's not very supportive of life. On top of that, Saturn's day cycle is super short. After Jupiter, Saturn has the second shortest day cycle, topping at only 11 Earth hours, making up one day! That's a short day! Although humans would have to figure out how to live with 11-hour days, it does mean that all parts of Saturn receive sunlight pretty regularly.

Unfortunately, scientists and engineers haven't yet found a way to live long-term in Saturn's clouds. Like Jupiter, we end up focusing on Saturn's moons, particularly Enceladus and Titan. Did you know these facts about Saturn?

- **Planet Type: Gas Giant**
- **Distance From Earth: 8.52 AU**
- **Distance From the Sun: 9.5 AU**
- **Radius: 36,184 miles (9 times the size of Earth)**
- **Rotation (Solar Day): almost 11 Earth hours (22,891 miles/hour)**
- **Orbit Around the Sun (a Year): 29.45 Earth years**
- **Moons: 78 (most important: Titan, Enceladus, and Phoebe)**

While other planets have rings, none of them have rings like Saturn. Saturn's rings are the greatest in the solar system, spreading out for 175,000 miles in orbit around the planet. How wide is that? Well, it would be the same as going around the Earth's equator seven times!

As the Dragon's Lance soars close to Saturn's rings, you might notice that Saturn's rings are made of rock and ice. This is why the rings shine so brightly! The variety of the size and shape of the rubble that creates Saturn's rings might surprise you. Most of the particles are somewhere between small dust particles and pieces of rock the size of houses. Only a few stones in the rings are as large as mountains. However, not all of the rings are the same. There are different hoops of rings, all of them moving at different speeds with different thicknesses and sometimes with gaps. Overall, the sight is beautiful. As we zoom close, we can see the rings themselves are usually only about 30 feet thick, vertically speaking. Speeding along with the rings in orbit around Saturn, we wonder whether the halo will collapse inward or one day just fly off into space.

The Dragon's Lance is going to land on Enceladus. Look out of the window to your right, and you will notice that this moon looks a little bit familiar. Like Europa, Enceladus has a surface with many lines and cracks on it. That is because underneath its icy surface lies a hidden ocean. We know that water lies beneath the thick ice because sometimes it sprays out into space through cracks.

Since Enceladus is around -330 °F, it is too dangerous to leave the ship without our EMU suits. After a short hike around the Dragon's Lance, we can relax and warm up in the Observatory. Through the massive windows, we can sit back and watch water geysers shoot out into the atmosphere. The Observatory will be nice and warm, protecting us from the piercing cold temperatures of Enceladus. Still, I think the planet looks pretty. It glimmers under the faint sunlight like a pearl.

Thanks to Enceladus's water and gas fountains, the particles form into ice and bind to rocks. As Enceladus orbits Saturn, its icy spray helps create Saturn's furthest ring, the E ring. Oh, look! As you can see, water just rose and arced out to space. Since there isn't a lot of gravity on Enceladus and no atmosphere, the water and gas shoot out into space at around 800 miles per hour, forming ice. I'm glad that happened during our visit.

Far below, warmth and internal heat must be keeping the water liquid. Otherwise, this planet would be frozen solid. If there is heat and water deep down, perhaps there are some simple life forms.

Scientists haven't yet figured out how to explore the ocean depths of Enceladus, but one day we might find out the answer to the mystery of Enceladus and discover new forms of alien life!

However, Enceladus isn't Saturn's only trending planet. Titan is a fascinating sphere because it is the only moon with weather systems like that on Earth! Tomorrow, we will head out from Enceladus to check out the storms on Titan.

As the second-largest moon in the solar system, Titan, like Ganymede, is larger than Mercury. Setting it apart from most other moons, Titan experiences weather systems thanks to its atmosphere and liquid methane. As we fly close to Titan, you may notice that it is a golden yellow, kind of like an egg yolk. That's because it has thick, dense clouds of nitrogen and methane. Here, sunlight is very faint, so once again, this planet is going to be cold. From the safety of our ship, we can observe Titan.

Since it is locked to Saturn's orbit and tilts about the same as Saturn, Titan experiences seasons similar to Saturn's. Each season will last around seven years. Experiencing Titan's seasonal weather would be great, but we can't take the Dragon's Lance down. Instead, we will rely on the Observatory to take a closer look. The human species wouldn't be able to settle on Titan. This moon has an enormous challenge within its atmosphere and structure.

Although Titan may have a core like Earth and an outer shell of ice, the surface consists of hydrocarbon grains and rivers of methane and ethane. As we get closer, you will be able to catch glimpses of Titan's land and water. Look out of that window! Wow!

We seem to be passing over heavy clouds. These clouds of nitrogen and methane would kill us if we breathed it in . . . or lit a match. That is why we are keeping our distance. Now, some rain is falling, but this isn't your everyday Earth rain. It's methane rain! Methane, a natural gas found on Earth, is everywhere on Titan. If we could figure out ways to transport this methane to Earth, we'd have an alternative source of fuel. However, too much methane in the atmosphere isn't good for Earth. Perhaps there are life forms on Titan who enjoy this methane and nitrogen gas world. Perhaps there are hidden life forms in the potential underground oceans below Titan's crust. Until more research happens, we will never know.

Named after the Roman god of agriculture and riches, Saturn is the father of Jupiter. Although ancient peoples of Earth knew and recognized Saturn in the night sky from early on, Galileo discovered Saturn's rings in the 1600s when he started to look at Saturn through a telescope. Since then, more rings and more moons have been observed orbiting Saturn. Today, with the help of binoculars, you might be able to glimpse its yellow-white twinkle in the night sky. If you have a good telescope, you'd be able to see its rings as well!

Look for Saturn in the evening from early August to the first week of January. It is exceptionally bright in the first week of August. You can find

Saturn in the mornings from February to August. Stargazers will usually find Saturn within the bounds of a constellation, depending on the year and the positioning of constellations. For example, in 2021, it appears close to Capricornus.

Snapping a unique selfie with Saturn might be strikingly cool, but wait until you reach the frozen worlds of Uranus and Neptune. These planets will take forever to appear, but they are certainly worth the journey! Let's kick back and chill while we speed toward the murky mysteries of Uranus. You have to think about what Uranus will be like.

The seventh planet, this light blue sphere, is quite distant from the Sun, making it super cold. Uranus probably consists of a dense core with a large outer shell of icy materials like water, ammonia, and methane. These form a syrup-like liquid that is quite hot. At the heart of this icy giant, temperatures might reach 9,000 °F. Above Uranus, there's a thick gaseous atmosphere of methane, hydrogen, and helium. This atmosphere is colder, often dipping as low as 375 °F. Sometimes, water and ammonia escape into the atmosphere.

Thanks to its dual nature of being both very hot and very cold, Uranus doesn't seem like it would support life. With little to no landmass along with intense pressure, landing on Uranus will be impossible. Breathing in the atmosphere would kill you instantly, and you would be dissolved in an instant, thanks to the harsh chemical mixtures.

With a 17-hour day, the cycle on Uranus is pretty short. However, some odd things are going on with Uranus. Look! We are getting closer. As the Dragon's Lance enters orbit around Uranus, you might notice some weird stuff going on. First off, Uranus doesn't rotate the same way as other planets. Like Venus, it is in retrograde—and its equator is 97° off. We can see how the rings don't circle the middle of the planet horizontally but circle around the planet's edge vertically. It certainly gives Uranus a beautiful silhouette. It also means that the faint sunlight shines on the planet's poles instead of the equator. Then there is the lopsided magnetic field. All of these factors make for some bizarre weather.

With winds that swirl at 560 miles an hour, Uranus experiences crazy weather storms. On various occasions, telescopes and probes have revealed that Uranus has white spots like Jupiter. This is a sign that intense storms are happening on the sphere. Usually, storms result from the interaction of heat and water in a planet's atmosphere. Thanks to the storms on Uranus, we can guess that there must be some insane heat levels generated by its unique rotation under those thick blue clouds.

If we could somehow land on Uranus, we'd be crushed instantly. However, overall, the mass of Uranus and the pull of gravity aren't as huge. On Uranus, a 100-pound person would only weigh around 89 pounds! That's not so terrible, right? Too bad that the weather and seasons on Uranus are so brutal. A year on Uranus lasts around 84 Earth years, which means that for 21 years straight, one side of the planet is plunged into darkness. Combine that with crazy winds and a toxic atmosphere, and you can see why most scientists look to the small moons and captured asteroids of Uranus as alternative resting points for space exploration. Let's find out more about Uranus!

- **Planet Type: Ice Giant**
- **Distance From Earth: 18.21 AU**
- **Distance From the Sun: 19.8 AU**
- **Radius: 15,759 miles (4 times the size of Earth)**

- **Rotation (Solar Day): 17 Earth hours (9,192.5 miles/hour)**
- **Orbit Around the Sun (a Year): 84 Earth years • Moons: 27**

Compared to Saturn, Uranus has fragile rings. They are so faint that you can't see them easily. Thanks to Voyager 2, we have better data about Uranus and its rings. Uranus has two sets. One set is closer, darker, and narrower; the other set is further out, with a blue and red coloration. There are quite a few ring belts around Uranus, but they are faint and more challenging to see. Fine dust surrounds some of the larger rings.

Although the rings on Uranus are smaller and thinner, it's still a very awe-inspiring experience to witness. Since they orbit Uranus's equator, they don't move around Uranus from east to west but flow up and around the blue planet, going north to south. Today, scientists believe that Uranus collided with a large object, an asteroid, or a minor object, which forced it to tilt, including its rings. Whatever the case, Uranus remains a mysterious, unique planet.

In 1781, William Herschel discovered Uranus with his telescope. Although he wanted to name it after King George III, the planet was named after the Greek god of the sky. I think that fits it better, don't you? Especially since Uranus is such a brilliant blue! Uranus is hard to find in the sky due to its color and distance from the Earth.

With good eyesight and a dark sky free of clouds, you can view Uranus from Earth. However, you will find it more easily with the help of binoculars. Want to see its rings? Get a good telescope, and you can view those rings from Earth! Like Saturn, Uranus hangs close to constellations. For example, in 2021, it remains close to Aries, appearing in the night sky between November and the first week of April. From May to November, you can find it in the early morning hours. It shines the brightest blue between August and December.

Uranus might have a unique spin on things, but we must leave it behind to get to Neptune. As the last planet in the Solar System, Neptune is a very distant planet and takes forever to reach. Out here, in the far reaches of space, the Sun appears very small. To find this planet, we will have to rely on technology instead of light.

We can finally get close to Neptune thanks to the Dragon Lance's incredible technology and our understanding of the solar system. This deep blue planet isn't very bright, but its colors are incredibly vivid. Like Uranus, Neptune is an ice giant, which means that it might have a small rocky core surrounded by superheated fluids of water, methane, and ammonia. Neptune is the densest gas giant, but scientists haven't been able to figure this planet out. This is because reaching Neptune is troublesome; nobody can enter the atmosphere to find out what lies under Neptune's thick clouds.

We know that way out here, far from the Sun, things get very cold. Furthermore, Neptune's days might be pretty short, lasting only 16 Earth hours, but its years are very long. One year on Neptune takes 165 Earth years! Since it is tilted on its axis, Neptune experiences seasons, but each season lasts 40 years. I can't imagine such a long winter.

In terms of pressure, Neptune is very dangerous as you get closer to its core. If it didn't have such tremendous strength, its gases would drift off. That doesn't mean Neptune is high in mass, though. The force of gravity on Neptune is moderate. If you weighed 100 pounds on Earth, you'd only be 13 pounds heavier on Neptune! However, primarily made of gas and icy liquids, a Neptune landing is certainly an impossibility.

Still, we can let the Dragon's Lance swoop close, so we can look at the winds rushing around the planet. Neptune's winds are the fastest in the solar system. They move three times as fast as the winds on Jupiter! The winds whirl by at 1,200 miles per hour. Neptune does experience storms that sometimes form into "mega dark spots thanks to this wild wind." In 1989, a "Great Dark Spot" appeared on the surface of the clouds. It looked to be about as large as the Earth itself! That's one storm you don't want to get caught in!

Even more dangerous, Neptune's atmosphere essentially holds hydrogen and helium, just like the other gas planets. There's a bit of methane in there as well, which makes Neptune look blue. Breathing in that air wouldn't be great. Between the toxic atmosphere, the cold temperatures, and the high winds, it's no surprise to learn that Neptune is nearly impossible for humans to settle on. Instead, we will hang out on Triton, one of Neptune's moons, and view the planet from there! Let's review some primary data about this mysterious planet!

- **Planet Type: Ice Giant**
- **Distance From Earth: 29.09 AU**
- **Distance From the Sun: 30.1 AU**
- **Radius: 15,299 (4 times the size of Earth)**
- **Rotation (Solar Day): 16 Earth hours (6,039 miles/hour)**

- **Orbit Around the Sun (a Year): 164.81 Earth years**
- **Moons: 14**

Neptune might not have as many moons as some other planets, but one of its moons has caught the eyes of astronomers. Triton, Neptune's largest moon, probably emerged from the Kuiper Belt and was captured by Neptune's gravity. One of the reasons why scientists believe it is a captured moon is that Triton has a retrograde orbit. It is the only minor member that orbits in the opposite direction of Neptune's rotation.

Triton has a rock and metal core covered by an icy mantle of frozen nitrogen. It is the only icy nitrogen moon in the system. It is quite a dense moon, but its atmosphere is surprisingly thin. The little atmosphere it has is mostly nitrogen with some methane. Triton's volcanic activity forms this atmosphere.

As we approach Triton, take notice of the far side of the crater. Take a look at those land formations over there. Those point to lava activity on Triton's surface—icy lava flows. Sometimes geysers spray out into the atmosphere, which Voyager 2 was able to record. Due to this volcanic activity, Triton is a rather exciting place. Look out of that window. Make sure to keep your camera ready! There goes a geyser! On Triton, it's freezing, so we will be staying inside the Observatory to watch the dynamic landscape of Triton and the serene view of Neptune.

Named after the Roman god of the sea, Neptune is a beautiful blue. Just like the Roman god, Neptune is gorgeous and dangerous. Too bad it can't be easily discovered in the night sky. Because its orbit is so far out, it takes 165 Earth years to revolve around the Sun one time. With good binoculars or a telescope, you can catch Neptune from mid-September to late February on a dark, clear night. From late March through September, you can find Neptune in the morning. It shines brightest between July and early November, but you will still need some visual aids for this mysterious planet.

Is the ride over? Here at the far reaches of the solar system, you might think so, but there is still so much to see! Let's swing by a few more awesome spots in our solar system, starting with the dwarf planets!

CHAPTER 8

THE MINOR MEMBERS

F ar out on the solar system's edge, the Sun seems much smaller than before, and its light is very distant and barely warms the cold dwarf planets and asteroids hovering on the edges of the solar system. Beyond, shrouded in darkness, ancient belts of rock and ice swirl around the system—the mysterious Kuiper Belt and Oort Cloud, which we have yet to observe and analyze fully.

Our tour shuttle has to rocket far from the solar system to even come close to Pluto or the Kuiper Belt. While we wait to arrive on Pluto, let's talk about its history. Pluto was considered a planet for a long time, but in 2006, the International Astronomical Union decided that Pluto wasn't a planet and demoted it to "dwarf planet" status. This is because while Pluto has moons, it shares its orbit and space with many other objects. To many people, however, Pluto remains their favorite. It isn't a well-known dwarf planet, so scientists still have a lot to learn about Pluto.

As we approach, we notice that it is indeed tiny for a planet. It is gray and brown, formed from rock and water ice. Many scientists theorize that Pluto may have emerged from the Kuiper Belt. Today, its wide, 248 Earth-year orbit around the Sun still takes it close to the Kuiper Belt, where sunlight takes forever to reach. Look where the Dragon's Lance is headed. Drawing close to Pluto, we can see it is covered with craters, mountains, and valleys. Pluto also appears to be covered in ice. Far off in the distance, it looks like glaciers have formed.

We are now landing close to the Burney Crater edge, where we can look over the great crater. It is full of smaller cavities, showing that Meteorites and other impacts have pounded Pluto. The size and lack of atmosphere do not protect it well, so the surface of Pluto shows all of its scars.

Of course, given how far off Pluto is, you can agree that it is frigid. It's almost -400 °F outside! This Observatory will keep us warm, but given how desolate Pluto is, work on all of the facilities has not been completed yet. We will stay here for a day to catch our breath and take time to look at the fringes of the Solar System.

Our travel schedule won't allow us to stay for a full Plutonian day. Because this planet rotates so slowly on its axis that it takes over 153 Earth hours for a single day to happen! You might be noticing something familiar! Like Venus and Uranus, Pluto has a retrograde rotation, which means that the sun rises and sets opposite to what we are used to on Earth.

Have you tried out the Indoor Stunt Bike Track? Here on Pluto, gravity is weak, just like on the Moon. So, if you weigh 100 pounds on Earth, you'll only weigh seven pounds on Pluto. Try jumping! Wow. You are certainly taking your time coming down. That's why you should try out the Bike Track. Grab a ticket and line up to try out the Course. Maybe you'll invent a new stunt out there!

- **Planet Type: Dwarf Planet**
- **Distance From Earth: 38 AU**
- **Distance From the Sun: 39 AU**
- **Radius: 715 miles**
- **Rotation (Solar Day): 143 Earth hours (10,600 miles/hour)**
- **Orbit Around the Sun (a Year): 248 Earth years**
- **Moons: 5**

There are many other dwarf planets in the solar system: Eris, Haumea, Makemake, and Ceres are other great examples of this category. However, not all of them live in the outer reaches of the solar system. You can find Ceres right in the asteroid belt. Many other dwarf planets might be hiding on the fringes of our solar system, but it will take a lot more time to find them all! Meanwhile, we can get a closer look at some asteroids and swing by a comet as well while we're at it.

When we passed through the asteroid belt, it might have been a bit of a letdown. Seeing as how far apart most asteroids are, flying through the belt isn't that exciting. Today, though, we can get up close and personal with an asteroid. Let's check out 99942 Apophis—it's 1,000 feet big. At first, scientists were worried that it would hit the Earth and an impact of that size would be hazardous. However, thanks to math and observation, scientists were able to analyze the trajectory of Apophis, showing that it won't hit Earth.

Look out of your left window. Now you can see Apophis. It's a jumbo piece of rock—around the size of almost four football fields. Made entirely of nickel and iron, it is stony with no atmosphere and no gravity—no potential for life. However, there is potential for future mining! Some asteroids hold up to 100 trillion dollars worth of gold, platinum, and other mineral resources. Asteroid mining is a future endeavor that will rival the California Gold Rush!

There are over one million asteroids in our solar system. Most of them are like Apophis, unable to support life. Asteroids don't have an atmosphere, much gravity, seasons, or weather. As a result, living life in the asteroid belt would be like living on a space station. Unlike other planets, asteroids might be ejected more easily from the belt and start a new orbit that brings them closer to the Sun or reorganizes them to the solar system's outer limits. Still, the study of asteroids continues today. Asteroids, after billions of years, hold secrets about how the solar system may have been formed.

Sometimes an asteroid breaks up or ejects a smaller asteroid, which then flies toward Earth and becomes a meteorite, falling through Earth's atmosphere. In 1940, a meteorite called Semarkona landed. After observing and analyzing the meteorite's characteristics, scientists realized that primitive asteroids and meteorites appear to have frozen magnetic fields trapped within. By measuring and analyzing these fragments, which are called chondrites, scientists can potentially find dust and debris that were melted into the chondrites and then frozen and preserved with magnetic fields intact. With these discoveries, scientists can make better guesses about the composition of the solar system while it was forming.

One of the easiest ways for astronomers and scientists to study asteroids is on Earth! Meteoroids—space rubble that includes anything from space dust to minor asteroids—might be too far away from Earth to study efficiently. However, if they get too close to a planet like Earth, possibilities open up!

As the meteoroid hits Earth's atmosphere, it begins to burn up, creating a short silvery trail. This is when the meteoroid changes into a meteor. It will burn up quickly if it is undersized, but it will take longer to burn up if it is more substantial, creating a longer silvery tail. Depending on where you live, you might be able to view a meteor shower once a year. Often, these meteors come from comets that pass by. Since the matter is ejected from a comet, the meteoroids are pretty small. Most of them are no bigger than a pea. On super-rare occasions, the Moon and Mars have been bombarded, causing the rock to split off from the planets and fall to Earth.

What if a meteoroid was truly big? Well, then, it might survive the planet's atmosphere and impact the terrain! Scientists call meteoroids that slam into a planet "meteorites." However, thanks to the strength of the Earth's atmosphere, by the time a large meteoroid hits the ground, it is only the size of a fist at most. If we had no atmosphere like the Moon, the meteoroid might not disintegrate and cause more damage on impact. Even with Earth's protective layers, meteoroids still often survive the fall to Earth, either impacting the ground with craters or exploding in the sky.

Today, scientists still search for meteorites as they look for more clues about how the solar system was formed and what changes have happened in our space neighborhood over time. Another important solar system also shares a lot of information with scientists: comets!

Since ancient times, comets have been very special. Ancient civilizations made up stories about them and observed them. Over time, humans were able to trace the cycles of comets that passed by Earth. However, we are still trying to figure out where comets come from. The leading theory suggests that comets are ejected from the Kuiper Belt or the Oort Cloud.

At first, far away from the Sun, a comet might look like an icy rock with some glowing gases. However, as the comet gets closer to the Sun, it warms up, and the gases expand. As a result, two long tails form—a dust tail and a gas tail—seen together and are called a coma. This bright, shining tail can get as long as 100,000 miles!

Although they are beautiful, the life of a comet is fragile and easily ended. If a comet gets too close to the fire of the Sun or the gravity of Jupiter, it may end up evaporating, crashing, or simply breaking up into pieces. On top of that, the orbits of comets can be quite long. Comets ejected from the Oort Cloud usually have a 200-year orbit or longer, while many comets from the Kuiper Belt have shorter orbits and can be seen more routinely. Do you know any famous comets? Comet Halley (also known as Halley's Comet) is only seen from Earth once every 76 years. Keep an eye out for Comet Halley in 2061!

Since we are so far from the Sun and Earth, we can explore further to look toward the farthest fringes of our solar system. Beyond this point, only a few instruments have analyzed what floats in the darkness. Although the Kuiper Belt was only a theory at first, the observations from NASA's New Horizons team have revealed distant objects that support the Kuiper Belt theory. Using the Hubble Space Telescope and the New Horizon's remote-controlled spacecraft, scientists will continue discovering more space objects beyond Pluto.

The Dragon's Lance will pause in its journey at this point, and we will gently drift along the edge of the quiet Kuiper Belt. From here, we can use the Observatory's super-powerful telescopes to look into the Kuiper Belt even more closely than we would be able to on Earth. We can see that we are already on the inner edge of the Kuiper Belt. It starts close to Neptune, around 30 AU from the Sun, and extends over 1,000 AU.

Pluto isn't alone in the Kuiper Belt. There are other dwarf planets, asteroids, comets, and space rubble. Some of the space rubble is quite big! Now, pause right there. Look through the telescope. Right there, it looks like a snowman, right? That's Arrokoth, the ancient snowman of the Kuiper Belt.

Arrokoth is a very primitive piece of red rock. It kind of looks like a snowman, but a floating one. This ancient snowman is very, very old: possibly around four billion years old! It was discovered by the New Horizons spacecraft in 2014. With the last of its fuel, it was able to take pictures of this ancient Kuiper Belt object. From observation and data analysis, we now know that objects like Arrokoth have rotation, and they orbit around the Sun. However, with no atmosphere or gravity, and no heat from the Sun, space objects like Arrokoth are not a place to live. Still, scientists believe there is evidence that Arrokoth and other Kuiper Belt objects may have different mixtures of materials. Scientists speculate that Arrokoth is composed of methanol, organic molecules, and water ice.

Thanks to the discovery of Arrokoth, we now have some compelling ideas of how planetesimals were formed long ago: the first steps to creating planets. Scientists may debate over the origin of Arrokoth, but most believe that scientific evidence points toward the cloud collapse theory. This means that Arrokoth came together slowly and relatively gently, not impacting each other at high speeds. Arrokoth, the ancient snowman, may have more information to share with us. We will have to revisit him sometime soon!

Oh! I have received many requests if we can continue to the Oort Cloud. Well, that's still difficult, even today. You might think you can catch a glimmer of it from the faint starlight that reaches around it, but no one has seen it yet. The Oort Cloud is so far away that we can never reach it in a lifetime at current speeds. From our Sun, it begins around 2,000 AU and probably ends around 5,000 AU. Right now, NASA's Voyager 1 is headed to the Oort Cloud, but it

won't get there for another 300 years . . . and it's traveling a million miles a day!

Now, if we could fly at the speed of light, we could reach the inner edges of the Oort Cloud within one month of flying at lightspeed. However, it would take us a total of almost eight light-years to get to the outer edges of the Oort Cloud. That's a thick cloud!

When you think of clouds, you might visualize soft, white, puffy clouds or thick, dark thunderstorm clouds. The Oort Cloud is made up of space rubble—dust, pebbles, and huge rocks. Some of the stones in the Oort Cloud are as large as mountains. Many of these might have frozen material or gases trapped inside. If one of them is knocked out of the Oort Cloud, it might impact the Kuiper Belt objects and become an asteroid. In some rare cases, it could speed toward the solar system and become a comet. These comets have 200-year-long orbits with very different angles from the other comets in the solar system, so they are harder to track. Still, astronomers are working on ways to reach and observe the Oort Cloud since there might be even more clues about how and when the solar system formed.

That's it for tonight! Tomorrow morning, we will use the Observatory telescopes to scan the Milky Way galaxy. For now, we can dream about what we might find.

CHAPTER 9

STARRY, STARRY NIGHTS

Today, we will do some Milky Way gazing. It might be surprising to you, but it's easier to see many of the Milky Way's stars and nebulae than it is to see the Oort Cloud. That's primarily due to light. We can see many beautiful sights thanks to the light of our galaxy's stars.

I'll start by explaining what the Milky Way is. On a clear night, you'll notice a soft white haze. It's much easier if there is no moon, no clouds, and no light pollution from the city. You see the Milky Way galaxy easier in the winter, thanks to shorter days and extended periods of darkness. The moon is hidden for long periods of time, causing the sky to have less light. However, you have a better chance of seeing the center of the Milky Way galaxy between March and September if you live in the Northern Hemisphere. As you look up, you will see the sky filled with stars and a blue-purplish haze that looks like soft clouds in space. That's the Milky Way galaxy!

The Milky Way is the galaxy our solar system orbits inside. We will talk more about that later on, but for now, it's meaningful to realize that this impressive spiral galaxy is home to billions of stars. Many of these stars have their own solar systems and planets. However, not all of the stars look the same. Just like we grow up and change according to the different cycles of our life, similarly, so too do stars transform over time. Let's have a look at a few stars and see what they can tell us about the life cycle of a star.

Let's zoom in on a beautiful scene. Across the black of space, purple and orange clouds of gas swirl together in mesmerizing shapes. As you look closer, you'll begin to see patterns similar to when you go cloud gazing on Earth. Is it a hand? Is it a person? Is it a bird? This is the Eagle Nebula, Messier 16, known as the Pillars of Creation. Here, stars are born from gas and dust. There are shining clouds (emission nebulae) and shadow-forming clouds (dark nebulae). Together, they show the different stages of a star's birth.

Nebulae are the cradles of stars. Due to gravity and matter ejected from other exploding stars, dust and gas form and collapse on each other. The star will begin to form a protostar thanks to ejected matter, gravity, and heat. This early stage of the star is unstable. There is a lot of gas clumped together with radiation, and it may start heating up. However, these disks of gas and dust only have a short time frame before they transform.

Check out the Orion Nebula. Move the telescope slowly . . . there! Isn't the Orion Nebula stunning? It looks like a swirl of pink and purple cotton candy. In fact, the Orion Nebula is another major factory for star birth. In the Orion Nebula, you will see a thick cluster of stars called the Trapezium Cluster. This is a big collection of stars. Scientists believe thousands of young stars may be hidden within the dense, cloudy cradle of the Orion Nebula.

Have a look at this. To the left, a little more, a little more… Right there. It's quite a giant sphere. Though, it isn't a planet. It's the brown dwarf Gliese 229 (GL229B). In 1995, the Hubble Space Telescope took photographs of Gliese. Next to its binary star companion, a red giant, this brown dwarf looks quite small. However, it is pretty big, definitely bigger than Jupiter. I can tell you what happened to Gliese 229.

Well, protostars don't have a long time to mature. Astronomers theorize that stars like our Sun remain in the protostar stage for around 100,000 years. However, that's not a lot of time in terms of the star's life cycle. If the protostar can't gain enough mass, it collapses into a brown dwarf. Brown dwarfs are too big to be planets and too small to be suns. So far, all brown dwarfs discovered usually orbit around a giant sun. Perhaps the more extensive sun got all of the mass. Either way, brown dwarfs are considered "failed stars."

On the other hand, stars that gain enough mass will grow in size and gain a healthy red-orange color. Hydrogen atoms fuse into helium and start nuclear fusion, entering the star into the main sequence phase. Even stars that achieve the main-sequence phase can be of different sizes.

Check out this star—Westerlund 1-26. This star is so big that if it sat in the middle of our solar system, its upper layers (the photosphere) would cover Saturn. That's a big star! In comparison, our Sun is quite small, but due to its size and efficiency, it will continue to burn for 10 billion years. Other more prominent stars, like Westerlund 1-26, may not last as long. After all, even the biggest and brightest must come to an end. Eventually, the star begins to expand into the first stages of its death: a red giant.

Red giants. Supergiants. Hypergiants: All of these star giants in our Milky Way galaxy! It makes you wonder what is really happening to these stars. Let's take a closer look at this faint golden-red glow in the night sky. It's Aldebaran, which is found in the constellation of Taurus. Around 65 light-years from Earth, this marvelous red giant star doesn't look like it's getting old, but it is.

Aldebaran and other red giants shine brilliantly because they have grown larger. Not all red giants act or look the same. Depending on their type, they may become red supergiants or vivid blue hypergiants before they become red giants. As the red giants begin to slow down in nuclear fusion, they instead start to heat up and expand. If the Sun turned into a red supergiant, it would grow, swallowing up many of the planets in the system. More than likely, however, our Sun will transform into a red giant, which will still burn up the Earth's atmosphere and render it unlivable.

As its nuclear fusion slows down to a stop, the stars grow, but they also lessen slowly in hotness. Finally, gravity begins to crush the star inward. Depending on the initial size of the star, there are different endings.

For smaller stars similar to our Sun, red giants collapse inward to become white dwarfs. ZTF J190132.9+145808.7 is a star that collapsed into a white dwarf around the size of our Moon. Many other white dwarfs tend to be much smaller. Look closely: The sun is very pale and very white. It isn't easy to see, is it? However, the mass of these stars is eight times the mass of our Sun. One teaspoon of white dwarf matter would weigh around 5.5 tons—that's about the weight of almost three elephants!

Eventually, the white dwarf will lose all heat and become a black dwarf, which will be hard to find. So far, there are no known black dwarfs in the Milky Way because our universe isn't old enough yet.

There are other ways for stars to die. Most of them involve fierce explosions and beautiful arrays of gases and radiation. For example, when a big star goes out with a bang, it can create a supernova, exploding its material up to 25,000 miles an hour. This will often help feed the birth of new stars. Humans have observed many supernovae from Earth over the centuries, with Chinese and Roman astronomers taking excellent notes of exploding stars. However, supernovae fade over time as the gases and particles drift through space.

A supernova isn't the only phase of a star's death. After their supernova, many stars collapse into super-small neutron stars. Like white dwarfs, the mass of these stars is super heavy—one teaspoon of the matter is around a billion tons. That would weigh as much as a mountain! Since gravity and pressure are so intense, these stars begin to emit strong magnetic fields and radiation. Many of these stars will spin around, shooting large pulses of energy and radiation outward. These are often called magnetars and pulsars.

Magnetars have crazy energy waves created from pressure on the star's magnetic fields within their crust. Recently, we have caught sight of a pulsar. Would you like to see one? Here, look through my telescope. Be extra observant. It's an intense ray of light whirling around and around like the beacon on a lighthouse. Except that isn't just light—it's also radiation. It's moving so fast! This is why this form of a neutron star is called a pulsar. It's pulsing, powerful beams of light and spraying out radiation particles.

I'm excited to show you something crazier and scarier than a pulsar. What happens when two neutron stars collide? You can only imagine what the explosion would look like! Let me find the newest kilonova we discovered recently. Take a look for yourself!

As you can see, there is a bright white center where the two stars have hit each other. From this explosion, great plumes of gases and radiation are shot out. After the kilonova calms down, a magnetar may remain. Still, it would be a dangerous neighborhood to hang out in!

I have one last thing to show you. Did we save the best for last? You let me know. Have a look through this telescope over here. What do you see? A large black circle surrounded by a ring of light. Is it a solar eclipse? No, it's a black hole!

As you know, after stars go supernova, they may transform into neutron stars. However, some stars continue to collapse under their own mass and gravity, turning into a black hole. Pulling on the space around them, black holes draw in all forms of matter—space rubble, planets, stars, and even whole galaxies! They even swallow all the light, or so it does appear that way, since no light can penetrate a black hole. However, scientists still aim to observe, analyze, and theorize about black holes and how they work. Perhaps black holes are just the universe's way of recycling matter. With additional time, we may be able to understand more. For now, we have to be careful because a black hole could consume this Observatory or even the Earth terribly quick!

Thanks to advancements in science and technology, we are getting better at observing space and all of its phenomena. We are even getting glimpses of the greater cosmos beyond. The last leg of our tour will be one final stop, where we will be able to peer beyond our galaxy. On the way, let's talk about all the incredible leaps in science and technology that have helped us get to where we are today.

CHAPTER 10

ASTRONOMICAL ADVENTURING

We are on our way to the last stop, but it will take some time. Let's visit the onboard virtual museum and explore the important events and artifacts that led to space exploration. We are going to have to go back several hundred years. After all, our interest in space has been going on for thousands of years. When humans began to notice the movement of the Sun, the Moon, and the stars.

Today, most of us take our calendars for granted, but in ancient times, watching the Sun and Moon while counting the days of the season was crucial for survival. Farming and agriculture could develop with knowledge of the seasons, leading to increased culture and civilization. However, the naked eye can only notice so much, even on a dark, clear night. The invention of the telescope started to change that.

Although nobody knows exactly when the telescope was invented, it is clear that glass and lenses were being developed as early as the 10th century. Over the years, Greek, Arab, and Italian scholars worked on eyeglasses and lenses, culminating in the first telescopes by the 1500s. In 1608, a Dutch inventor in the Netherlands filed the first patent application for a telescope. However, by then, telescopes were in use as a way to navigate the ocean.

It wasn't until the 1600s, when Galileo Galilei started to recognize the uses of the telescope, that they became famous. Not only did telescopes have value for navigators, but also for the army and academics. Galileo began to point the telescope upward, noticing that this enhanced his view of the sky. Galileo used the device to identify four planets, including Jupiter and Saturn. He also concluded that the Earth revolved around the Sun—a heavily criticized idea not welcomed at the time.

As time went on, future innovations in the telescope would allow astronomers and scientists to see farther and better. The telescopes became longer and more complicated. Isaac Newton used Kepler's writings to build one of the first reflecting telescopes that used a combination of lenses and mirrors to gain clear images.

Since then, many kinds of telescopes have continued to develop in size and power used by professionals and amateurs. The Hubble Space Telescope launched into orbit around Earth in 1990. With its clear view of the skies, the Hubble Space Telescope has helped contribute to many findings in the solar system and the galaxy. Other telescopes are being developed to replace Hubble with newer orbital and deep space telescopes. Some of these telescopes will hover further away from Earth and, with modern technology, discover more answers about the formation of stars and galaxies.

Eventually, just looking at the sky wasn't good enough for astronomers, and it was time to get a closer look at the solar system. However, while rockets were expensive and dangerous, space probes were a great short-term solution for scientists. Scientists would better understand Earth and its neighbors with the data sent back from space probes. Between the former Soviet Union (Russia) and the United States, they launched various instruments from the 1950s onward. The first probe, Sputnik 1, was launched in 1957 by the Soviet Union. The following year, the United States sent its probe, Explorer 1, out into space. This was the start of the Space Race, which lasted for a few decades.

Over the years, space agencies have sent many tools to investigate outer space. Each of them was able to tell us a little bit more about the solar system's planets, asteroids, and other space objects. One of the most famous probes has flown very far—Voyager 1. Leaving Earth in 1977, it slowly traveled, passing by Jupiter and Saturn before leaving for the edge of our solar system. Since then, it's still flying onward, carrying messages, instructions, and information from Earth, including Earth languages and music samples. In July 2022, Voyager 1 was 157 AU from the Sun, and Voyager 2 was 130 AU from the Sun. Voyager 1 is traveling 38 thousand mph and gaining 3.5 AU per year. As these two probes travel forward, they take readings of magnetic fields, low-energy charged particles, cosmic rays, and other investigations with plasma waves.

More recently, in 2006, NASA launched the New Horizons probe to investigate Pluto and the Kuiper Belt. Since then, it has returned fascinating discoveries, such as Pluto and its moons, and the discovery of Kuiper Belt objects like Arrokoth. What else will the New Horizons probe find? Only time will tell.

Simply looking at or receiving data wasn't giving scientists enough information, so they invented spacecraft with landing capabilities to allow for more detailed investigations. Check out this little robot—it's the Lunokhod Moon Robot. Two of them traveled over the Moon, sampling the soil and measuring its properties. The Apollo program had similar space rovers, which also explored the Moon.

All of this was pretty exciting, but most scientists were already thinking about Mars. Any probes to Venus were crushed within a few hours, so it was clear that Mars was the better option for exploration and human settlement. By the late 1990s, engineers built new rovers for Mars exploration. In 1997, Sojourner was the first rover to land on Mars. And then later models of robotic rovers, Spirit and Opportunity, continued searching for life in 2004. Although many rovers are built for shorter operation periods, Opportunity lived for 15 years before finally shutting down in 2019.

A new rover has since landed on Mars. Curiosity, the latest Mars rover, landed in Gale Crater to observe and transmit data about Mars's geological history. Check out the 3D model of Curiosity. As you will see, the design allows the robot to move over rough terrain strategically. Its six wheels carry a boxy body that holds the computers and instrumentation used to explore and observe Mars. Scientists can study the data sent to Earth by Curiosity, equipped with drills and cameras. Rovers will continue to have a crucial role in space exploration.

Living in space was another natural step for humankind. Living in orbit would help support the orbital telescopes and allow for new experiments that scientists could only do in space. As a result, both Russia and America worked on developing various space stations that could orbit Earth long-term.

In the early 1970s, America launched the Skylab space station with the intention that it would remain in orbit for a decade. Though by 1977, it became clear that Skylab was coming out of orbit and would re-enter Earth's atmosphere. It ended up crashing into the southeastern Indian Ocean and a section of Western Australia. The

Russians also set up various space stations in the 1970s, but the last generation of Russian space stations, Mir, was retired in 2001.

By the late 1990s, it became clear that a proper long-term space station had to come from a team effort between countries. Canada, Japan, the EU, Russia, and Brazil joined with America's NASA to help build the International Space Station. Each country provided equipment, technology, or modules to add to the station over time. In 1998, two parts of the station were joined in orbit, and since then, other pieces and modules have been added. Since 2000, the ISS has always had astronauts visiting and living there. Even when NASA retired, SpaceX and other programs continued to use the ISS as a place for experimentation and research.

Between 2010 and 2016, China also launched two test space stations in preparation for its official Chinese space station. In 2021, China launched the first section of the Tiangong Space Station. Tiangong, which means "heavenly palace" in Mandarin, aims to promote scientific investigation in space, continuing the legacy of astronauts, cosmonauts, and taikonauts into the future.

Do you think that living in orbit was enough for humans? Of course not! The former Soviet Union and the United States started to figure out ways to send humans into orbit and further to the Moon. It all began long ago when rocketry was first invented. When Chinese people shot fireworks into the sky long ago, they didn't know their discovery would change history. However, over time, the British army began to use this gunpowder for more effective rocket weaponry in war. Over time, rockets became sizable and more powerful.

Three men changed everything: Konstantin E. Tsiolkovsky, Robert Goddard, and Hermann Oberth. Each of their studies into rocketry, particularly around the early 1900s and during the World Wars, helped develop rocketry so that large rockets could start to carry humans. The first rockets carried probes and animals into space. Then, in 1961, the USSR launched Vostok, which shot the first human into orbit. On board, Yuri Gagarin made history by reaching orbit and completing one complete circle of Earth. Vostok is famous for carrying the first woman, Valentina Tereshkova, into space in 1963.

During the same year, the United States began the Mercury program with astronaut Alan Shepard flying the Freedom 7. The Mercury program didn't get an American astronaut into orbit until 1962 with John Glenn in the Friendship 7. For the next few decades, the former Soviet Union cosmonauts and American astronauts competed to push new boundaries in space. In 1965, Voskhod 2 made history with one of its cosmonauts, Alexei Leonov, achieving the first spacewalk.

Even after the Cold War and the Space Race ended, Russia and America pursued advancement in skill and technology to one day put someone on the moon. In 1968, the Apollo program began, and it aimed at rethinking how to achieve spaceflight and spacecraft engineering. These iterations eventually landed the first two humans on the moon in 1969. When Neil Armstrong and Buzz Aldrin set foot on the moon, they conducted experiments, stored samples, and explored some of the Moon's surface.

After Apollo's last flight in 1972, NASA paused spaceflight for nearly a decade, returning in 1981 with a redesigned, reusable space shuttle. Challenger and Columbia, two of the five shuttles, were destroyed in tragic accidents, but the Endeavour and other shuttles continued flying until they were retired in 2011. During this time, many advancements were achieved, such as the creation of the International Space Station. Even after the USA discontinued the Apollo program, humans were not giving up on space. By now, other countries, like Japan, Russia, and China, have started collaborating. Other engineers, scientists, and astronauts from Canada, India, and the EU have helped with innovative technology and rocketry to enhance space flight, exploration, and living.

Since 2003, China has begun to launch its private space vehicles, known as the Shenzhou. Carrying Chinese taikonauts to space, these space flights have helped solidify China's involvement in space, supporting their independent space stations. NASA has also continued to assess new ways to reach outer space, proposing the new Space Launch System (SLS), planning to accomplish a return mission to the Moon. This long, slim spacecraft has only three parts: two solid rocket boosters and a heavy rocket launch system, making up the main spacecraft. The Orion spacecraft, where the astronauts will live, is attached to the top of the main rocket. Like other rocket styles, this multiple-launch system hopes to provide reusable systems to gain orbit and visit the Moon.

Meanwhile, other non-national companies decided to try their hand at space travel. SpaceX, a commercial crew program, began to send astronauts to space in the Crew Dragon capsules. In 2020, SpaceX's Dragon successfully brought four astronauts to the ISS and then returned them to Earth six months later. With each successful mission accomplished, SpaceX worked to prepare for the next phase: space tourism. As space travel slowly moves away from national and scientific interests, more corporations may look to space as a new frontier for business and entertainment. None of these exciting new opportunities would have happened if it weren't for the brave men and women who worked to push the boundaries of knowledge.

Well, that was a lovely day of astronomy history. Now we have arrived at the last observatory. From here, we can see further beyond distant galaxies and alien stars with their unknown possibilities for life. Life on another planet might look different from what we expect, though. Before we get excited about aliens, let's look at how scientists aim to study alien life.

Usually, the search for alien life ties to our current understanding of where we came from and how the universe evolved the way it did. Astrobiology is a field of study where life in the universe is observed. Thanks to technology and scientific advancements, astronomers, biologists, chemists, geologists, and engineers can work together to find ways that humans can live on other planets. Scientists can also theorize about what kind of biological life might exist in the universe. They look to answer questions like "Where did life come from?", "What kind of planet could support life?" or "Can we live on other planets besides Earth?"

As you can see, astrobiology is trying to answer many questions. Some of these questions are not just about ideas or knowledge: Some are looking at ways to live on other planets. Since this field of study has beneficial aspects, it relies on data from telescopes, probes, rovers, and missions to learn more about the origins of the universe, the evolution of our solar system and planet, along with the potential for life on other planets.

Of course, you might be thinking of aliens right away. What do you think aliens would look like? Are they tall and thin? Are they gray or green? Do they have intelligence or not? Will they be able to communicate with us? Movies and video games have a lot of different ideas about what alien life might look like. However, scientists are also interested in microbial life. The presence of even single-cell organisms on a planet would be a significant step toward understanding the universe and how it evolved.

This is why the search for exoplanets is significant. Exoplanets are any planets we find outside of our solar system. Most of them, like our solar system, orbit their star. However, on occasion, some planets aren't bound by a star's gravity and just float through the galaxy. These types are called rogue planets.

Since planets don't shine like stars do, finding them has always been a challenge; however, there are various methods to find exoplanets today. It would be nice to get a photo of an exoplanet, but it's tough. Using direct imaging (photography with a telescope) is only possible when the planets are massive (like Jupiter). Around 300 light-years from Earth, two giant exoplanets orbit TYC 8998-760-1. They weren't able to be seen easily, but the European Southern Observatory (ESO) took pictures of them through the very high-powered telescope named SPHERE. More popular ways to find exoplanets are through the use of indirect observation. For example, when a giant planet crosses in front of the Sun, it shows up as a shadowy dot. In reviewing the transmission of light, scientific instruments might notice a blip on the radar. This shows that a planet had passed in front of the Sun. Another method is to watch the gravitational field of stars. Although stars are more powerful, planets still can cause a slight wobble in the star's gravitational field. When scientists notice a significant effect on the star, they can know that there may be one or more planets in the system. Usually, a variety of methods are used together to finalize the discovery of an exoplanet.

Still, just the existence of an exoplanet doesn't mean that life might prevail on the planet. If the world is too close to the Sun or too far away, life as we know it might struggle to survive or have already been destroyed. As a result, besides finding exoplanets, scientists and astrobiologists look for exoplanets within the habitable Goldilocks Zone. Planets in this zone have a greater chance of developing life, even if it is only a microscopic single-cell organism. Whatever alien life looks like, the quest hasn't ended. You, too, could be part of the search for life in the universe! There is a lot of work to get done, especially considering how massive our universe is!

CHAPTER 11

TO INFINITY AND BEYOND!

Now that we know the long history and changing goals of space exploration, it's time to look further and see even more! We have safely arrived at our final destination. From this vantage point, we can look beyond our Milky Way out into intergalactic space.

Oh! What's this photograph? You might think it's a black painting covered in white, yellow, blue, and pinkish splatter, but it is the deepest and sharpest infrared image of intergalactic space. All of these specks are galaxies! Each galaxy has its exclusive collection of stars and exoplanets! This is the famous First Deep Field image by the James Webb Space Telescope (JWST). Overflowing with detail and excitement. In the picture is a galaxy cluster named SMACS 0723. As an iconic photograph of the universe, it makes you realize that the universe is so vast, our minds—even with the help of mathematics—can not comprehend it.

Pushing at the limits of human knowledge, scientists are scanning the universe for wisdom about the origin of life as we know it. Further beyond our galaxy, we begin to understand the big picture of the cosmos. Suddenly, our massive Sun and distant planets feel very small as we come to realize that our solar system sits on a spur of a galaxy that is only one of the millions of other galaxies. Where do we even start? We will begin by looking at our home galaxy.

Let's put on these VR goggles and access the most powerful telescopes to zoom around the Milky Way and the universe. Floating above the Milky Way, we can see its spiral shape. It's overwhelming how big it is, right? Still, astronomers are already trying to map our home galaxy. Just like any organism, astronomers have begun to name the different parts of our Milky Way galaxy. Let's zoom in on the center.

In the middle, where the galaxy shines brightest, the Galactic Bar is located. In this area, the stars are densely clustered. Here, in the galaxy center, the stars in the bar cluster together as they rotate around the center of the galaxy. However, scientists have noticed that dark matter appears to be slowing the galaxy's rotation down over time.

Although we often look at pictures of the Milky Way Galaxy from the top down, you may have noticed that it isn't flat. Look further up in the telescope. You can see that while the galaxy is thinner on the edges, it gets fatter toward the middle. Closer to the center, you see more globular clusters hanging above and below the universe. Each of these clusters can hold around 10 million stars! Many scientists believe that globular clusters are usually more symmetrical, with more stars in the middle and fewer stars around them. That makes them look circular in shape. As a result, globular clusters are quite old. They have moved out quite a ways from the galaxy's center, and their orbits take forever to complete—sometimes a few hundred million years! Compared to globular clusters, newer open clusters of stars are the places where new stars are maturing. That's where our Sun is found.

From the Galactic Bar, two major arms extend and bend around. They are called Perseus and Scutum-Centaurus. Two other, less-defined arms, Sagittarius and Norma, are located between the two primary arms. Between the Perseus and Sagittarius arms, a smaller half-arm sticks out. This is called the Orion Arm, where our Sun is located. It isn't too close to the galaxy's center, which is excellent because a black hole hangs out at the center of each universe.

Let's fly back to the center of the galaxy. All you see is a bright white glare. That's because the concentration of stars here is so intense. You might remember that black holes are black. So where is it?

Many theories suggest that the Milky Way, unlike some other galaxies, has an inactive black hole. Scientists wonder if we have dark matter at the center of our galaxy. If the Milky Way has a black hole at its center, it is unique. Our black hole isn't greedy, so it hasn't grown much. However, there are other galaxies out there with active black holes. Sometimes the active galaxies go into overdrive. They swirl around their hungry black holes. As entire stars and solar systems are eaten, massive amounts of energy are released around the black holes. These huge jet streams of energy and radiation rush out from the top and bottom. These jet streams can impact other galaxies as well.

Follow me: I'll show you a fantastic sight of an active black hole. There! What do you see? It's very faint, but you can see a glimpse of a quasar with special imaging. These images combine various ray and radiation data to create a picture for us. The puffy cloud is the center of the galaxy. Around it, the disk of stars is swirling. However, from the middle, you can see red streams of yellow-white heat and energy. These are the jet streams of a quasar. If it were closer to us and pointed at us, our galaxy could be impacted by particles and radiation. It turns out that our universe is just as dangerous as our galaxy! This only increases the importance of studying galaxies. The more we know, the more we can recognize the signs of disturbance in our own galaxy.

When scientists look at galaxies, they can consider the shapes and locations of galaxies. With these kinds of observations, astronomers can begin to map the universe itself!

Galaxy shapes are pretty straightforward. There are three main shapes: spiral, elliptical, and irregular. The Milky Way is an example of a spiral galaxy. These galaxies usually have a spinning disk with arms that twirl around. The center has a slight bulge, but the outer edges are thinner. Compared to the Milky Way, some spiral galaxies have longer or bigger bars in the center. Take a look at NGC 1672. It only has a couple of arms, but that's because the center bar is quite long. Still, you see a spiral, sort of. The second type of galaxy, elliptical, is older. They look like stretched-out circles with a slight bulge in the middle. Since they have less gas and dust, ellipticals don't create new stars as much as spiral galaxies do. However, there are many different sizes of elliptical galaxies. One of the massive galaxies in the universe, M87, is an elliptical galaxy. The final category, irregular galaxies, is rarer. Many of them aren't rounded or spiraled. Many of them don't have arms or bulges in the center. Scientists believe that irregular galaxies are formed when a galaxy bumps into another galaxy. Do you want to see what an irregular galaxy looks like? Check out NGC 1472A. It kind of looks like a nebula, right? It's indeed more vast than a nebula and stunningly consists of billions of stars!

As mentioned, galaxies can bump into other galaxies. This means that the location of a galaxy is also fundamental to astronomers. Do galaxies travel alone? As a matter of fact, scientists have discovered that galaxies are likely to travel in packs. These are called clusters. Clusters range in size from 10 galaxies to 10,000! Of course, they are probably drawn together due to gravitational pull. That's not all. These clusters draw to each other in different ways, forming superclusters.

Imagine it. Our Milky Way has some close neighbors: the Andromeda Galaxy, the Large Magellanic Cloud, and the Small Magellanic Cloud, among others. Would you like to have a look at our galactic neighbors? Near Australia and the Polynesian islands, the native inhabitants noticed these clouds long ago. Today, we would have to fly to Australia, New Zealand, or a Polynesian island to view the Magellanic clouds. Thankfully, out here in space, we can look around more easily. Let's take a peek at the Larger Magellanic Cloud. In VR, the Magellanic clouds appear clearly and vividly. On Earth, they look like a faint blue, gray, and pinkish swirling haze. Although it shines a bit, it seems at first glance like a nebula. However, the Large and Small Magellanic Clouds are galaxies that orbit the Milky Way. Scientists believe that these clouds are slowly moving away from the Milky Way so that they won't be crashing into us any time soon.

The Andromeda Galaxy, the Magellanic Clouds, and our Milky Way, alongside other galaxies, form the Local Group, a small cluster of galaxies. However, this whole group is considered a "poor" group because there are less than 1,000 galaxies in the collection. The Virgo cluster, with over 2,000 galaxies, is considered moderately "rich." The Coma cluster holds around 10,000 galaxies!

If galaxies are attracted to each other, it makes sense then that clusters are also attracted to each other and form superclusters. The Local Group, where our galaxy is located, forms a supercluster with the Virgo Cluster and a few others. Our entire cosmos is filled with superclusters of all sizes. Some of these are massive, such as the Coma Supercluster or the Perseus-Pisces Supercluster.

Unfortunately, our glimpse of the universe so far has been limited. The Hubble Space Telescope revealed more than we ever imagined and is now being replaced by better, newer telescopes that can see more than just a single window into the galaxy. I know it's exciting to think about all those worlds that haven't yet been discovered. However, until humans figure out how to cross such vast distances, we will never be able to reach a supercluster, much less another galaxy. Let's get the Orion Tour crew and passengers back onto the Dragon's Lance. It's time to head home!

As we approach Earth, you may breathe a sigh of relief. The trip was super fun, but there is something special about our Earth. Here, we feel truly at home on this small planet, orbiting an insignificant Sun within an insignificant galaxy. The air, the sunlight, the moonlight, and the diverse forms of nature around us help support human life. Earth, for now, is a one-of-a-kind.

Oh! That's the announcement! The waiting period is over. Everyone is stashing their food and packing up the last of their belongings. It's time for one more zero-G somersault. As the shuttle begins to shift in its trajectory, we are getting ready to reenter the Earth's atmosphere. We'd better prepare and get into our suits for reentry!

At first, things might feel bumpy and even get a bit warm. You shouldn't be worried, though. You know now that this is the way Earth protects us from meteors. We'll land safe and sound thanks to our suits and this hi-tech cabin.

Where are we landing? Off the coast of Florida! Let me see. I'll check to ensure you are suited up and have the buckle attached correctly.

The cabin shaking is perfectly normal. Things start to heat up a bit. As a roaring and warmth fills the air, you catch a glimpse of the dark skies of space, giving way to the brilliant blue of Earth's atmosphere. Parachutes pop out, slowing the capsule down. As our speed decreases, gravity begins to take hold. That was an excellent entry. We've hit the ocean. As promised, you have safely arrived back home. The crew will open the door for you!

CONCLUSION

Thank you for joining Orion Tours for the epic journey around our solar system. Don't forget to have your *Cosmic Passport* stamped on your way out.

"What's that light doing in the sky? How do the eight planets differ? When do comets get their tails?" Now you might have a better idea of how to answer!

The wonders of the night sky have inspired humans for thousands of years, influencing art, music, and storytelling. From the star-studded tombs of the Egyptian mummies to more recent films, like *The Martian, Interstellar, and Gravity*, outer space has provided humans with a source of awe and respect. Both dangerous and beautiful, deep space has revealed itself as an opportunity for humans to grow through inspiration and challenge. Similarly, exploring space provides educational value and a sense of self-confidence and determination that can positively impact you and your family through life, learning teamwork and individual self-expression cultivated through exploration.

Don't forget your free Space Explorer Activity Pack! Go to my website: https://www.pantheonspace.com/solarkids

You're invited to connect with us on **Facebook, Instagram, and TikTok** by searching for @PantheonSpace. See you in the comments!

Websites

- NASA Kids' Club (www.nasa.gov/kidsclub/index.html)
- NASA Space Place (spaceplace.nasa.gov)
- NASA at Home (https://www.nasa.gov/nasa-at-home-for-kids-and-families)
- National Air and Space Museum (airandspace.si.edu)
- Jet Propulsion Laboratory (www.jpl.nasa.gov/edu/)

YouTube Channels

- NASA Goddard
- Hubble Space Telescope
- Fraser Cain
- NOVA
- Vsauce
- Veritasium

Other Resources

- (iPhone/Android) Solar System Scope
- (iPhone/Android) Audio Sky Tours - Astronomy
- (iPhone/Android) Star Walk Kids: Astronomy Game
- (Oculus Rift/Playstation VR) Apollo 11 VR
- (Mac/Windows/Linux) Kerbal Space Program

Launch Into Space!

The wonders of the night sky have inspired humans for thousands of years, influencing art, music, and storytelling. From the star-studded tombs of Egyptian mummies to more recent films, like *Apollo 13*, space has provided humans with a source of awe and respect. Both dangerous and beautiful, space has revealed itself to be an opportunity for humans to grow through inspiration and challenge. Similarly, your exploration of space provides not only educational value but also a sense of self-confidence and determination that can positively impact you and your family throughout life. Learn the teamwork and individual self-expression cultivated through exploration, and get ready to launch into life and space!

If you enjoyed this book or received value from it in any way, I appreciate it if you could take a moment to **leave a review for this book on Amazon.** As an independent author, it means the world to me! I read and consider all feedback. **Thank you for your time and support**.

REFERENCES

Absolute Cosmos. (2020). Why Do All the Planets Orbit More or Less in the Same Plane? [YouTube Video]. In *YouTube*. https://www.youtube.com/watch?v=Qv_C0ZXGuWk

American Museum of Natural History. (2013). *Formation of Our Solar System | AMNH.*

American Museum of Natural History. https://www.amnh.org/exhibitions/permanent/the-universe/planets/formation-of-our-solar-system

Anderberg, J. (2019, May 15). *15 Constellations Every Man Should Know (And How to Find Them).* The Art of Manliness. https://www.artofmanliness.com/articles/15-constellations-every-man-should-know/

Arizona State University. (2014a, November 14). *The measurements point to shock waves traveling through the cloud of dusty gas around the newborn sun as a major factor in solar system formation.*

Astronomy.com. https://astronomy.com/news/2014/11/magnetic-fields-frozen-into-meteorite-grains-tell-shocking-tale-ofsolar-system-birth

Arizona State University. (2014b, November 14). *The measurements point to shock waves traveling through the cloud of dusty gas around the newborn sun as a major factor in solar system formation.*

Astronomy.com. https://astronomy.com/news/2014/11/magnetic-fields-frozen-into-meteorite-grains-tell-shocking-tale-ofsolar-system-birth

Astronomy Notes. (2020). *Lives and Deaths of Stars.* Astronomynotes.com.

https://www.astronomynotes.com/evolutn/s3.htm

Bartels, M. (2020, May 23). *Here's every spaceship that's ever carried an astronaut into orbit.*

Space.com; Space. https://www.space.com/every-crewed-spacecraft-human-spaceflight-history.html

BBC. (2014a, June 20). *The Sky at Night - Section 1: The lunar seas.* BBC; BBC. https://www.bbc.co.uk/programmes/articles/5gdrKwHtXhRkcq0xHDdhqvj/section-1-the-lunar-seas

BBC. (2014b, June 20). *The Sky at Night - Section 1: The lunar seas.* BBC; BBC.

https://www.bbc.co.uk/programmes/articles/5gdrKwHtXhRkcq0xHDdhqvj/section-1-the-lunar-seas

BBC News. (2020, May 28). *Nasa SpaceX launch: Evolution of the spacesuit.* BBC News; BBC News. https://www.bbc.com/news/science-environment-52787365

Belloni, M. (2014). *Sidereal/Solar Day.* Compadre.org.

https://www.compadre.org/osp/EJSS/3567/7.htm

Betz, E. (2021, March 5). *The Kármán Line: Where does space begin?* Astronomy.com.

https://astronomy.com/news/2021/03/the-krmn-line-where-does-space-begin

Biezen, M. van. (2014). Astronomy - Ch. 2: Understanding the Night Sky (18 of 23) Solar Day vs Sideral Day [YouTube Video]. In *YouTube.* https://www.youtube.com/watch?v=kzmvGOhhEOM

Bright Hub. (2010, February 23). *What is the Average Asteroid Speed?* Brighthub.com.

https://www.brighthub.com/science/space/articles/64710/

Bright Side. (2019, September). *How Long the Human Body Could Survive on Each Planet of the Solar System.* BrightSide — Inspiration. Creativity. Wonder.

https://brightside.me/wonder-curiosities/how-long-the-human-body-could-survive-on-each-planet-of-th e-solar-system-794065/

Brown, H. (2016, March 2). *Good Question: How Many People Have Gone To Space?* Cbslocal.com;

WCCO | CBS Minnesota. https://minnesota.cbslocal.com/2016/03/02/good-question-astronauts/

Byrd, D. (2018, October 9). *Meet Proxima Centauri, closest star to sun | Astronomy Essentials | EarthSky.* EarthSky | Updates on Your Cosmos and World. https://earthsky.org/space/proxima-centauri-our-suns-nearest-neighbor/

Canadian Space Agency. (2015). *Dwarf planets.*

https://www.asc-csa.gc.ca/eng/astronomy/solar-system/dwarf-planets.asp

Cessna, A. (2009, July 27). *Planetesimals - Universe Today.* Universe Today.

https://www.universetoday.com/35974/planetesimals/#:~:text=A%20planetesimal%20is%20an%20objec

t,%2C%20rock%2C%20and%20other%20materials.&text=The%20term%20re
fers%20to%20small,the%2 0Russian%20astronomer%20Viktor%20Safronov.

Chemical Safety Facts. (2015, March 23). *Ammonia Uses and Benefits | Chemical Safety Facts.*

ChemicalSafetyFacts.org. https://www.chemicalsafetyfacts.org/ammonia/

Chen, M. (2016, October 26). *Kerbal Space Program.* Common Sense Education; Common Sense

Education. https://www.commonsense.org/education/game/kerbal-space-program

China National Space Administration. (2021). *China launches first section of its massive space station.* Cnsa.gov.cn.
http://www.cnsa.gov.cn/english/n6465652/n6465653/c6811970/content.html

Choi, C. Q. (2021, June 9). *Earth's sun: Facts about the sun's age, size and history.* Space.com; Space. https://www.space.com/58-the-sun-formation-facts-and-characteristics.html

Chow, D. (2011, October 19). *The Universe: Big Bang to Now in 10 Easy Steps.* Space.com; Space. https://www.space.com/13320-big-bang-universe-10-steps-explainer.html

Cofield, C. (2016, August 24). *How We Could Visit the Possibly Earth-Like Planet Proxima b.*

Space.com; Space. https://www.space.com/33844-proxima-b-exoplanet-interstellar-mission.html

Collins, G., & Johnson, T. V. (2014). Ganymede and Callisto. *Encyclopedia of the Solar System,* 813–829. https://doi.org/10.1016/b978-0-12-415845-0.00037-2

Cox, L. (2017, December 21). *Who Invented the Telescope?* Space.com; Space.

https://www.space.com/21950-who-invented-the-telescope.html#:~:text=The%20first%20person%20to %20apply,with%20a%20convex%20objective%20lens.

Crane, L. (2021). *Mars's surface hosts millions of towering dust devils every day.* New Scientist.

https://www.newscientist.com/article/2143217-marss-surface-hosts-millions-of-towering-dust-devils-eve ry-day/

Crowley, C. (2020, November 17). *Hubble sees the brightest kilonova yet | Space | EarthSky.*

EarthSky | Updates on Your Cosmos and World. https://earthsky.org/space/unusual-kilonova-infrared-light-neutron-stars-grb/

Dorrian, G., & Whittaker, I. (2021, July 5). *Tiangong: astronauts are working on China's new space station – here's what to expect.* The Conversation.

https://theconversation.com/tiangong-astronauts-are-working-on-chinas-new-space-station-heres-what-t o-expect-163375

Dunlop, J. (2020, April 14). *How to Find the Milky Way at Night.* ExpertPhotography; ExpertPhotography. https://expertphotography.com/how-to-find-the-milky-way-at-night/

EarthSky. (2013, March 24). *Surprises in Venus' south pole vortex on EarthSky | Space | EarthSky.*

EarthSky | Updates on Your Cosmos and World. https://earthsky.org/space/surprises-in-venus-south-polar-vortex/

EarthSky. (2020a, January 5). *What is the asteroid belt? | Space.* EarthSky.

https://earthsky.org/space/what-is-the-asteroid-belt/

EarthSky. (2020b, January 16). *How far is Betelgeuse? | Astronomy Essentials.* EarthSky | Updates on

Your Cosmos and World. https://earthsky.org/astronomy-essentials/how-far-is-betelgeuse/

EarthSky. (2020c, July 26). *1st direct image of 2 giant exoplanets orbiting a sunlike star | Space | EarthSky.* EarthSky | Updates on Your Cosmos and World. https://earthsky.org/space/1st-photos-gas-giant-exoplanets-tyc-8998-760-1/

EarthSky. (2020d, November 12). *What is a supernova? | Astronomy Essentials | EarthSky.*

EarthSky | Updates on Your Cosmos and World. https://earthsky.org/astronomy-essentials/definition-what-is-a-supernova/

EarthSky. (2020e, December 9). *What is a coronal mass ejection?* EarthSky | Updates on Your Cosmos and World. https://earthsky.org/space/what-are-coronal-mass-ejections/

EducationalAppStore Ltd. (2018). *10 of the Best Space Apps for Kids | Educational App Store.*

Educational App Store. https://www.educationalappstore.com/app/category/astronomy-apps

Emerging Technology from the arXiv. (2018, June 22). *This is how many people we'd have to send to Proxima Centauri to make sure someone actually arrives.* MIT Technology Review; MIT Technology Review. https://www.technologyreview.com/2018/06/22/142160/this-is-how-many-people-wed-have-to-send-to -proxima-centauri-to-make-sure-someone-actually/

ESA. (2013). *What is a "launch window"?* Esa.int. https://www.esa.int/Science_Exploration/Space_Science/What_is_a_launch_window#:~:text=The%20l aunch%20window%20is%20a,some%20point%20in%20the%20future.

Espenak, F. (2020). *Solar Eclipses for Beginners.* Mreclipse.com. http://www.mreclipse.com/Special/SEprimer.html

European Southern Observatory. (2021). *ESO.* Eso.org. https://www.eso.org/public/outreach/eduoff/vt-2004/mt-2003/mt-mercury-history.html

Exoplanet Exploration. (2021, April 2). *Overview | What is an Exoplanet? – Exoplanet Exploration:*

Planets Beyond our Solar System. Exoplanet Exploration: Planets beyond Our Solar System. https://exoplanets.nasa.gov/what-is-an-exoplanet/overview/

Exploratorium. (2021). *Your Weight on Other Worlds.* Exploratorium.edu. https://www.exploratorium.edu/ronh/weight/

Fox, A. (2021, July 2). *Incredibly Dense White Dwarf Star Packs the Mass of the Sun Into the Size of the Moon.* Smithsonian Magazine; Smithsonian Magazine. https://www.smithsonianmag.com/smart-news/incredibly-dense-white-dwarf-star-packs-mass-sun-sizemoon-180978107/

Freudenrich, C. (2000, December 14). *How Space Suits Work.* HowStuffWorks. https://science.howstuffworks.com/space-suit.htm

Georgiou, A. (2020, May 21). *Why Did the Space Shuttle Program End?* Newsweek; Newsweek. https://www.newsweek.com/why-space-shuttle-program-end-1505594

Gorrell, D. (1970a). *Britannica Kids: Solar System.* Common Sense Education; Common Sense

Education. https://www.commonsense.org/education/app/britannica-kids-solar-system

Gorrell, D. (1970b). *Star Walk Kids: Astronomy Game.* Common Sense Education; Common

Sense Education. https://www.commonsense.org/education/app/star-walk-kids-astronomy-game

GreekMythology.com. (2018, March 13). *Icarus - Greek Mythology.* Greekmythology.com; GreekMythology.com. https://www.greekmythology.com/Myths/Mortals/Icarus/icarus.html

Grossman, L. (2019, January 29). *There's a snowman in space.* Science News for Students. https://www.sciencenewsforstudents.org/article/theres-snowman-space

Gunn, D. A. (2019). *Do all planets have magnetic fields?* BBC Science Focus Magazine; BBC

Science Focus Magazine. https://www.sciencefocus.com/space/do-all-planets-have-magnetic-fields/

Harland, D. M. (2021). space station | Definition, History, & Facts | Britannica. In *Encyclopædia Britannica.* https://www.britannica.com/technology/space-station

Hathaway, D. H. (n.d.). *A Scale Model of the Solar System.* Retrieved June 29, 2021, from https://solarscience.msfc.nasa.gov/papers/hathadh/SolarSystemModel.pdf

Howell, E. (2015a, February 11). *What Are The Most Famous Stars? - Universe Today.* Universe Today. https://www.universetoday.com/45775/famous-stars/

Howell, E. (2015b, July 22). *Globular Clusters: Dense Groups of Stars.* Space.com; Space.

https://www.space.com/29717-globular-clusters.html

Howell, E. (2017, September 20). *Halley's Comet: Facts About the Most Famous Comet.* Space.com;

Space. https://www.space.com/19878-halleys-comet.html

Howell, E. (2018, February 8). *International Space Station: Facts, History & Tracking*. Space.com; Space. https://www.space.com/16748-international-space-station.html

Howell, E. (2019a, January 9). *What Are Elliptical Galaxies?* Space.com; Space.

https://www.space.com/22395-elliptical-galaxies.html

Howell, E. (2019b, March 25). *Venera 13 and the Mission to Reach Venus*. Space.com; Space.

https://www.space.com/18551-venera-13.html

Howell, E. (2021, May 17). *What is the Big Bang Theory?* Space.com; Space.

https://www.space.com/25126-big-bang-theory.html

Imster, E. (2017, February 15). *What causes the aurora borealis? | Earth | EarthSky*. EarthSky | Updates on Your Cosmos and World. https://earthsky.org/earth/what-causes-the-aurora-borealis-or-northern-lights/

Langevin, S. (1970). *Apollo 11 VR*. Common Sense Education; Common Sense Education. https://www.commonsense.org/education/game/apollo-11-vr

Launius, R. (n.d.). *Space Stations: Base Camps to the Stars*. Retrieved July 8, 2021, from
https://repository.si.edu/bitstream/handle/10088/17610/nasm_Space_Stations-v35_Chapter_23rev-Fin al.pdf

Lunar and Planetary Institute. (2021). *Planetesimals*. Usra.edu.

https://www.lpi.usra.edu/education/timeline/gallery/slide_3.html

MacPhee, M. (2020a, July 29). *The awesome beauty of the Eagle Nebula | Astronomy Essentials | EarthSky*. EarthSky | Updates on Your Cosmos and World. https://earthsky.org/clusters-nebulae-galaxies/the-awesome-beauty-of-m16-the-eagle-nebula/

MacPhee, M. (2020b, July 29). *The awesome beauty of the Eagle Nebula | Astronomy Essentials | EarthSky*. EarthSky | Updates on Your Cosmos and World. https://earthsky.org/clusters-nebulae-galaxies/the-awesome-beauty-of-m16-the-eagle-nebula/

Mars Exploration Program. (2020). *NASA Mars Curiosity Rover*. NASA's Mars Exploration Program. https://mars.nasa.gov/msl/home/

McClure, B. (2017, October 23). *What is an astronomical unit? | Space | EarthSky.* EarthSky |

Updates on Your Cosmos and World. https://earthsky.org/space/what-is-the-astronomical-unit/

Mooney, M. (2020, September 30). *Olympus Mons | The Biggest Hotspot in the Solar System - Lowell Observatory.* Lowell Observatory.

https://lowell.edu/olympus-mons-the-biggest-hotspot-in-the-solar-system/#:~:text=Olympus%20Mons%20is%20a%20shield,of%20the%20state%20of%20Arizona).

NASA. (2011a). *Astronauts Tour Future White Room, Crew Access Tower.* NASA.

https://www.nasa.gov/feature/astronauts-tour-future-white-room-crew-access-tower

NASA. (2011b). *Layers of the Sun.* NASA.
https://www.nasa.gov/mission_pages/iris/multimedia/layerzoo.html

NASA. (2011c). *Mars Rovers Overview.* NASA.
https://www.nasa.gov/mission_pages/mer/overview/index.html

NASA. (2011d). *Mars Rovers Overview.* NASA.

https://www.nasa.gov/mission_pages/mer/overview/index.html

NASA. (2011e). *NASA - Yuri Gagarin: First Man in Space.* Nasa.gov.

https://www.nasa.gov/mission_pages/shuttle/sts1/gagarin_anniversary.html

NASA. (2011f). *NASA Knows: Rockets and Space Vehicles (Grades 5-8).* NASA.

https://www.nasa.gov/audience/forstudents/5-8/features/nasa-knows/vehicles/index.html

NASA. (2011g). *Ring-a-Round the Saturn.* NASA.

https://www.nasa.gov/audience/forstudents/k-4/stories/nasa-knows/ring-a-round-the-saturn.html

NASA. (2011h). *Scientists to Io: Your Volcanoes Are in the Wrong Place.* NASA.
https://www.nasa.gov/topics/solarsystem/features/io-volcanoes-displaced.html

NASA. (2011i). *Scientists Uncover Origins of Sun's Swirling Spicules.* NASA.

https://www.nasa.gov/feature/goddard/2017/scientists-uncover-origins-of-the-sun-s-swirling-spicules

NASA. (2011j). *The Apollo Missions*. NASA.
https://www.nasa.gov/mission_pages/apollo/missions/index.html

NASA. (2011k). *Valles Marineris: The Grand Canyon of Mars*. NASA.
https://www.nasa.gov/multimedia/imagegallery/image_feature_83.html

NASA. (2011l). *Venus Weather Not Boring After All, NASA/International Study Shows*. Nasa.gov.

https://www.nasa.gov/topics/solarsystem/features/venus-temp20110926.html

NASA. (2011m). *What is a solar prominence?* NASA.

https://www.nasa.gov/content/goddard/what-is-a-solar-prominence

NASA. (2014). *Three Ways to Travel at (Nearly) the Speed of Light*. NASA.

https://www.nasa.gov/feature/goddard/2019/three-ways-to-travel-at-nearly-the-speed-of-light

(2015a). *Eating in Space*. NASA.

https://www.nasa.gov/audience/foreducators/stem-on-station/ditl_eating

NASA. (2015b). *Eating in Space*. NASA.
https://www.nasa.gov/audience/foreducators/stem-on-station/ditl_eating

NASA. (2015c). *Free Time in Space*. NASA.

https://www.nasa.gov/audience/foreducators/stem-on-station/ditl_free_time

NASA. (2015d). *Morning Routine in Space*. NASA.

https://www.nasa.gov/audience/foreducators/stem-on-station/ditl_morning_routine

NASA. (2015e). *New Horizons*. NASA.
https://www.nasa.gov/mission_pages/newhorizons/main/index.html

NASA. (2015f). *Sleeping in Space*. NASA.
https://www.nasa.gov/audience/foreducators/stem-on-station/ditl_sleeping

NASA. (2015g). *What Is a Rocket?* NASA.
https://www.nasa.gov/audience/forstudents/5-8/features/nasa-knows/what-is-a-rocket-58.html

NASA. (2015h). *What Is a Spacesuit?* NASA.
https://www.nasa.gov/audience/forstudents/k-4/stories/nasa-knows/what-is-a-spacesuit-k4.html

NASA. (2015i). *What Is a Spacewalk?* NASA.

https://www.nasa.gov/audience/forstudents/k-4/stories/nasa-knows/what-is-a-spacewalk-k4.html

NASA. (2015j). *What Is the Hubble Space Telescope?* NASA.

https://www.nasa.gov/audience/forstudents/k-4/stories/nasa-knows/what-is-the-hubble-space-telecope-k 4.html

NASA. (2015k). *What Is the International Space Station?* NASA.
https://www.nasa.gov/audience/forstudents/k-4/stories/nasa-knows/what-is-the-iss-k4.html

NASA. (2016a). *Active Galaxies and Quasars - Introduction.* Nasa.gov.

https://imagine.gsfc.nasa.gov/science/objects/active_galaxies1.html#:~:text=Active%20galaxies%20are%2 0galaxies%20that,the%20rest%20of%20the%20galaxy.

NASA. (2016b). *Galaxies.* Nasa.gov.
https://imagine.gsfc.nasa.gov/science/objects/galaxies1.html

NASA. (2016c). *International Space Station Facts and Figures.* NASA.
https://www.nasa.gov/feature/facts-and-figures

NASA. (2016d). *International Space Station Facts and Figures.* NASA.

https://www.nasa.gov/feature/facts-and-figures

NASA. (2017). *Neutron Stars, Pulsars, and Magnetars - Introduction.* Nasa.gov.

https://imagine.gsfc.nasa.gov/science/objects/neutron_stars1.html

(2019a). *NASA Kids' Club.* NASA. https://www.nasa.gov/kidsclub/index.html

NASA. (2019b). *Space Launch System.* NASA.

https://www.nasa.gov/exploration/systems/sls/index.html

NASA. (2019c). *Spacesuit Basics.* NASA. https://www.nasa.gov/feature/spacewalk-spacesuit-basics

NASA. (2019d). *Spacesuit Basics.* NASA.
https://www.nasa.gov/feature/spacewalk-spacesuit-basics

NASA. (2019e, June 20). *5 Ways to Find a Planet | Explore – Exoplanet Exploration: Planets Beyond our Solar System*. Exoplanet Exploration: Planets beyond Our Solar System. https://exoplanets.nasa.gov/alien-worlds/ways-to-find-a-planet/

NASA. (2019f, October 16). *Earth's Atmosphere: A Multi-layered Cake*. Climate Change: Vital

Signs of the Planet. https://climate.nasa.gov/news/2919/earths-atmosphere-a-multi-layered-cake/

NASA. (2019g, December 19). *In Depth | Comets – NASA Solar System Exploration*. NASA Solar

System Exploration. https://solarsystem.nasa.gov/asteroids-comets-and-meteors/comets/in-depth/

NASA. (2020a). *How Long is a Year on Other Planets?* Nasa.gov.

https://spaceplace.nasa.gov/years-on-other-planets/en/

NASA. (2020b). *Imagine the Universe!* Nasa.gov. https://imagine.gsfc.nasa.gov/features/cosmic/nearest_star_info.html

NASA. (2020c). *Imagine the Universe!* Nasa.gov. https://imagine.gsfc.nasa.gov/features/cosmic/nearest_superclusters_info.html

NASA. (2020d). *Imagine the Universe!* Nasa.gov.

https://imagine.gsfc.nasa.gov/features/cosmic/local_supercluster_info.html

NASA. (2020e). *Milky Way's Black Hole's Flash Illuminated Gas Far Outside Our Galaxy*. NASA.

https://www.nasa.gov/feature/goddard/2020/intense-flash-from-milky-ways-black-hole-illuminated-gasfar-outside-of-our-galaxy

NASA. (2020f). *NASA: 60 Years & Counting - The Future*. NASA.

https://www.nasa.gov/specials/60counting/future.html

NASA. (2020g). *Solar System Sizes | NASA Solar System Exploration*. NASA Solar System Exploration. https://solarsystem.nasa.gov/resources/686/solar-system-sizes/

NASA. (2020h). *Spot The Station*. Nasa.gov. https://spotthestation.nasa.gov/

NASA. (2020i). *Spot The Station*. Nasa.gov. https://spotthestation.nasa.gov/

NASA. (2020j). *Voyager - Fast Facts*. Nasa.gov.

https://voyager.jpl.nasa.gov/frequently-asked-questions/fast-facts/#:~:text=As%20of%20April%202020%202C%20Voyager,about%203.3%20AU%20per%20year.

(2020k). *What Is the Space Launch System?* NASA.

https://www.nasa.gov/audience/forstudents/5-8/stories/nasa-knows/what-is-the-space-launch-system.ht ml

NASA. (2020l, September 16). *Sun 3D Model*. NASA Solar System Exploration. https://solarsystem.nasa.gov/solar-system/sun/in-depth/

NASA. (2020m, October 8). *Enceladus 3D Model*. NASA Solar System Exploration. https://solarsystem.nasa.gov/moons/saturn-moons/enceladus/in-depth/

NASA. (2020n, October 8). *Saturn 3D Model*. NASA Solar System Exploration. https://solarsystem.nasa.gov/planets/saturn/in-depth/

NASA. (2020o, October 12). *Europa 3D Model*. NASA Solar System Exploration. https://solarsystem.nasa.gov/moons/jupiter-moons/europa/in-depth/

NASA. (2020p, October 12). *Jupiter 3D Model*. NASA Solar System Exploration. https://solarsystem.nasa.gov/planets/jupiter/in-depth/#otp_structure

NASA. (2020q, October 12). *Pluto 3D Model*. NASA Solar System Exploration.

https://solarsystem.nasa.gov/planets/dwarf-planets/pluto/in-depth/

NASA. (2020r, November 10). *Io*. NASA Solar System Exploration.

https://solarsystem.nasa.gov/moons/jupiter-moons/io/overview/

NASA. (2020s, November 10). *Titan - Overview*. NASA Solar System Exploration.

https://solarsystem.nasa.gov/moons/saturn-moons/titan/overview/

NASA. (2020t, November 19). *Moons*. NASA Solar System Exploration.

https://solarsystem.nasa.gov/moons/overview/

NASA. (2020u). What You Need to Know About Saturn's Moon Titan [YouTube Video]. In *YouTube*. https://www.youtube.com/watch?v=lr4r70DWShk

NASA. (2021a). *Eyes On Exoplanets – Exoplanet Exploration: Planets Beyond our Solar System.* Nasa.gov. https://exoplanets.nasa.gov/eyes-on-exoplanets/#/planet/KELT-9_b/

NASA. (2021b). *Home | NASA Space Place – NASA Science for Kids.* Nasa.gov. https://spaceplace.nasa.gov/

NASA. (2021c). *Mars Exploration: Multimedia.* Nasa.gov. https://mars.nasa.gov/gallery/atlas/olympus-mons.html

NASA. (2021d). *NASA - Countdown 101.* Nasa.gov. https://www.nasa.gov/mission_pages/shuttle/launch/countdown101.html

NASA. (2021e). *NASA - Telescope History.* Nasa.gov. https://www.nasa.gov/audience/forstudents/9-12/features/telescope_feature_912.html

(2021f). *NASA - What Is a Space Probe?* Nasa.gov. https://www.nasa.gov/centers/jpl/education/spaceprobe-20100225.html

NASA. (2021g). *NASA - What Is a Space Probe?* Nasa.gov. https://www.nasa.gov/centers/jpl/education/spaceprobe-20100225.html

NASA. (2021h). *NASA - Solid Rocket Boosters.* Nasa.gov. https://www.nasa.gov/returntoflight/system/system_SRB.html

NASA. (2021i). *Telescope History.* Nasa.gov. https://www.nasa.gov/audience/forstudents/9-12/features/telescope_feature_912.html

NASA. (2021j). *The First Person on the Moon.* Nasa.gov. https://www.nasa.gov/audience/forstudents/k-4/stories/first-person-on-moon.html#:~:text=On%20July%2020%2C%201969%2C%20Neil,walked%20around%20for%20three%20hours.

NASA. (2021k). *The First Person on the Moon.* Nasa.gov. https://www.nasa.gov/audience/forstudents/k-4/stories/first-person-on-moon.html#:~:text=On%20July%2020%2C%201969%2C%20Neil,walked%20around%20for%20three%20hours.

NASA. (2021l). *What is a brown dwarf?* Nasa.gov.
https://starchild.gsfc.nasa.gov/docs/StarChild/questions/question62.html

NASA. (2021m). *What Is a Space Probe?* Nasa.gov.

https://www.nasa.gov/centers/jpl/education/spaceprobe-
20100225.html#:~:text=Probes%20send%20dat
a%20back%20to%20Earth%20for%20scientists%20to%20study.&text=Sputnik
%201%20was%20the%20fi
rst%20probe%20to%20go%20into%20space.&text=Once%20probes%20could
%20reach%20space,probe%20to%20study%20another%20planet.

NASA. (2021n, February 4). *Titan 3D Model.* NASA Solar System Exploration.
https://solarsystem.nasa.gov/moons/saturn-moons/titan/in-depth/

NASA. (2021o, February 4). *Triton 3D Model.* NASA Solar System Exploration.

https://solarsystem.nasa.gov/moons/neptune-moons/triton/in-depth/

NASA. (2021p, February 15). *Neptune.* NASA Solar System Exploration.
https://solarsystem.nasa.gov/planets/neptune/overview/

NASA. (2021q, February 15). *Uranus.* NASA Solar System Exploration.

https://solarsystem.nasa.gov/planets/uranus/overview/

NASA. (2021r, March 22). *Gas Giant | Planet Types – Exoplanet Exploration:
Planets Beyond our Solar System.* Exoplanet Exploration: Planets beyond Our Solar
System. https://exoplanets.nasa.gov/what-is-an-exoplanet/planet-types/gas-giant/

NASA. (2021s, June 15). *Ganymede 3D Model.* NASA Solar System Exploration.

https://solarsystem.nasa.gov/moons/jupiter-moons/ganymede/in-depth/

(2021t, June 25). *In Depth | Asteroids – NASA Solar System Exploration.* NASA
Solar

System Exploration. https://solarsystem.nasa.gov/asteroids-comets-and-
meteors/asteroids/in-depth/

NASA Goddard. (2021). *One Image, One Million Galaxies.* Www.youtube.com.

https://www.youtube.com/watch?v=u0ihatxXQFI

NASA Science. (2017). *The Milky Way Galaxy | NASA Solar System Exploration.*
NASA Solar System Exploration. https://solarsystem.nasa.gov/resources/285/the-
milky-way-galaxy/

NASA Science. (2018). *What Are Constellations? | NASA Space Place – NASA Science for Kids*. Nasa.gov. https://spaceplace.nasa.gov/constellations/en/

NASA Science. (2019a, June 6). *Oort Cloud*. NASA Solar System Exploration. https://solarsystem.nasa.gov/solar-system/oort-cloud/overview/

NASA Science. (2019b, December 19). *In Depth | Asteroids – NASA Solar System Exploration*.

NASA Solar System Exploration. https://solarsystem.nasa.gov/asteroids-comets-and-meteors/asteroids/in-depth/

NASA Science. (2019c, December 19). *In Depth | Asteroids, Comets & Meteors – NASA Solar System Exploration*. NASA Solar System Exploration. https://solarsystem.nasa.gov/asteroids-comets-and-meteors/in-depth/

NASA Science. (2019d, December 19). *In Depth | Beyond Our Solar System – NASA Solar System Exploration*. NASA Solar System Exploration. https://solarsystem.nasa.gov/solar-system/beyond/in-depth/

NASA Science. (2020a, October 12). *Earth with Clouds 3D Model*. NASA Solar System Exploration. https://solarsystem.nasa.gov/planets/earth/in-depth/

NASA Science. (2020b, October 12). *Jupiter 3D Model*. NASA Solar System Exploration. https://solarsystem.nasa.gov/planets/jupiter/in-depth/

NASA Science. (2021a). *Basics of Space Flight - Solar System Exploration: NASA Science*. NASA Solar System Exploration. https://solarsystem.nasa.gov/basics/chapter5-1/

NASA Science. (2021b, February 4). *Triton 3D Model*. NASA Solar System Exploration. https://solarsystem.nasa.gov/moons/neptune-moons/triton/in-depth/

NASA Science. (2021c, February 15). *Neptune*. NASA Solar System Exploration. https://solarsystem.nasa.gov/planets/neptune/overview/

NASA Science. (2021d, February 15). *Saturn*. NASA Solar System Exploration. https://solarsystem.nasa.gov/planets/saturn/overview/

NASA Science. (2021e, February 15). *Uranus*. NASA Solar System Exploration. https://solarsystem.nasa.gov/planets/uranus/overview/

NASA Science. (2021f, February 15). *Uranus*. NASA Solar System Exploration.

https://solarsystem.nasa.gov/planets/uranus/overview/

NASA Science. (2021g, February 17). *Beyond Our Solar System*. NASA Solar System Exploration.

https://solarsystem.nasa.gov/solar-system/beyond/overview/

NASA Science. (2021h, February 19). *Planets*. NASA Solar System Exploration.

https://solarsystem.nasa.gov/planets/overview/

NASA Science. (2021i, March 3). *Lunar Phases and Eclipses | Earth's Moon – NASA Solar System Exploration*. NASA Solar System Exploration. https://solarsystem.nasa.gov/moons/earths-moon/lunar-phases-and-eclipses/

NASA Science. (2021j, May 19). *Earth's Moon*. NASA Solar System Exploration. https://solarsystem.nasa.gov/moons/earths-moon/overview/

NASA Science. (2021k, May 19). *Mars 3D Model*. NASA Solar System Exploration.

https://solarsystem.nasa.gov/planets/mars/in-depth/

NASA Science. (2021l, June 11). *Mercury*. NASA Solar System Exploration. https://solarsystem.nasa.gov/planets/mercury/overview/

NASA Science. (2021m, June 11). *Venus*. NASA Solar System Exploration. https://solarsystem.nasa.gov/planets/venus/overview/

NASA Science. (2021n, June 14). *Kuiper Belt*. NASA Solar System Exploration.

https://solarsystem.nasa.gov/solar-system/kuiper-belt/overview/

NASA Solar System Exploration. (2018a, January 23). *Mars Moons*. NASA Solar System Exploration.

https://solarsystem.nasa.gov/moons/mars-moons/overview/?page=0&per_page=40&order=name+asc&search=&condition_1=6%3Aparent_id&condition_2=moon%3Abody_type%3Ai like&condition_3=mo on%3Abody_type

NASA Solar System Exploration. (2018b, September 25). *Enceladus*. NASA Solar System

Exploration. https://solarsystem.nasa.gov/missions/cassini/science/enceladus/

NASA Solar System Exploration. (2019a). *Mercury's Caloris Basin*. NASA Solar System Exploration. https://solarsystem.nasa.gov/resources/2266/mercurys-caloris-basin/

NASA Solar System Exploration. (2019b, November 13). *Overview | Arrokoth (2014 MU69)*.

NASA Solar System Exploration. https://solarsystem.nasa.gov/solar-system/kuiper-belt/arrokoth-2014-mu69/overview/

NASA Solar System Exploration. (2019c, December 19). *In Depth | Comets*. NASA Solar System Exploration. https://solarsystem.nasa.gov/asteroids-comets-and-meteors/comets/in-depth/

NASA Solar System Exploration. (2019d, December 19). *In Depth | Oort Cloud*. NASA Solar

System Exploration. https://solarsystem.nasa.gov/solar-system/oort-cloud/in-depth/

NASA Solar System Exploration. (2019e, December 19). *Our Solar System*. NASA Solar System

Exploration. https://solarsystem.nasa.gov/solar-system/our-solar-system/in-depth/

NASA Solar System Exploration. (2020a, April 23). *In Depth | 1P/Halley*. NASA Solar System

Exploration. https://solarsystem.nasa.gov/asteroids-comets-and-meteors/comets/1p-halley/in-depth/

NASA Solar System Exploration. (2020b, May 11). *In Depth | Meteors & Meteorites*. NASA Solar System Exploration. https://solarsystem.nasa.gov/asteroids-comets-and-meteors/meteors-and-meteorites/in-depth/

NASA Solar System Exploration. (2020c, July 20). *In Depth | Apophis*. NASA Solar System

Exploration. https://solarsystem.nasa.gov/asteroids-comets-and-meteors/asteroids/apophis/in-depth/

NASA Solar System Exploration. (2021a, June 9). *Overview | Cassini*. NASA Solar System Exploration. https://solarsystem.nasa.gov/missions/cassini/overview/

NASA Solar System Exploration. (2021b, June 14). *Kuiper Belt*. NASA Solar System Exploration. https://solarsystem.nasa.gov/solar-system/kuiper-belt/overview/

NASA Solar System Exploration. (2021c, June 25). *In Depth | Asteroids*. NASA Solar System Exploration. https://solarsystem.nasa.gov/asteroids-comets-and-meteors/asteroids/in-depth/#many_shapes_and_sizes _otp

NASA Space Place. (2020). *What Is Gravity?* Nasa.gov. https://spaceplace.nasa.gov/what-is-gravity/en/

NASA Space Place – NASA Science for Kids. (2016). *How Old Are Galaxies?* Nasa.gov. https://spaceplace.nasa.gov/galaxies-age/en/#:~:text=Astronomers%20believe%20that%20our%20own,about%20500%20million%20years%20ago.

NASA Space Place – NASA Science for Kids. (2018). *What Are Constellations?* Nasa.gov. https://spaceplace.nasa.gov/constellations/en/

NASA Space Place – NASA Science for Kids. (2019). *Sunspots and Solar Flares*. Nasa.gov. https://spaceplace.nasa.gov/solar-activity/en/

NASA Space Place – NASA Science for Kids. (2020a). *How Do We Weigh Planets?* Nasa.gov. https://spaceplace.nasa.gov/planets-weight/en/

NASA Space Place – NASA Science for Kids. (2020b). *What Is the Solar Cycle?* Nasa.gov. https://spaceplace.nasa.gov/solar-cycles/en/

NASA Video. (2013). Station Tour: Harmony, Tranquility, Unity [YouTube Video]. In *YouTube*. https://www.youtube.com/watch?v=ukws3oLMDc8

NASA/JPL Edu. (2019, November 19). *Educator Guide: Let's Go to Mars! Calculating Launch Windows | NASA/JPL Edu*. NASA/JPL Edu. https://www.jpl.nasa.gov/edu/teach/activity/lets-go-to-mars-calculating-launch-windows/

NASA's Hyperwall. (2018, May 15). *Hyperwall: Hubble Ultra Deep Field*. Nasa.gov. https://svs.gsfc.nasa.gov/30946

National Air and Space Museum. (2021a). *Homepage*. Si.edu. https://airandspace.si.edu/

National Air and Space Museum. (2021b). *Wind and Dust on Mars | Exploring the Planets*. Si.edu. https://airandspace.si.edu/exhibitions/exploring-the-planets/online/solar-system/mars/wind/#:~:text=Su rface%20winds%20typically%20move%20about,per%20hour%20during%20du st%20storms.

National Geographic. (2017, February 16). *7 Magical Places to View Auroras*. Travel; National Geographic. https://www.nationalgeographic.com/travel/article/7-aurora-destinations

National Geographic Society. (2011, December 13). *Planetary Size and Distance Comparison*.

National Geographic Society. https://www.nationalgeographic.org/activity/planetary-size-and-distance-comparison/

National Geographic Society. (2019, October 22). *Atmosphere*. National Geographic Society. https://www.nationalgeographic.org/encyclopedia/atmosphere-RL/

National Oceanic and Atmospheric Administration. (2021a). *Coronal Mass Ejections | NOAA / NWS Space Weather Prediction Center*. Noaa.gov. https://www.swpc.noaa.gov/phenomena/coronal-mass-ejections

National Oceanic and Atmospheric Administration. (2021b). *Planet Rotations Dataset | Science On a Sphere*. Noaa.gov. https://sos.noaa.gov/datasets/planet-rotations/

Nemiroff, R., & Bonnell, J. (2018). *APOD: 2018 May 27 - Coronal Rain on the Sun*. Nasa.gov. https://apod.nasa.gov/apod/ap180527.html

NOVA Online. (2021). *Runaway Universe | Galaxies, Clusters, and Superclusters*. Pbs.org. https://www.pbs.org/wgbh/nova/universe/tour_ggs.html

Owen, J. (2020). How did the Orbiter Vehicle work? (Space Shuttle) [YouTube Video]. In *YouTube*. https://www.youtube.com/watch?v=cFBRawYov00

Parks, J. (2019, January 2). *Ultima Thule: The cosmic snowman*. Astronomy.com. https://astronomy.com/news/2019/01/ultima-thule-the-cosmic-snowman

QuarkstoQuasars. (2014, January 15). *The Life Cycle Of A Star:* Futurism; Futurism. https://futurism.com/the-life-cycle-of-a-star

Rao, J. (2020, December 31). *When, where and how to see the planets in the 2021 night sky.* Space.com; Space. https://www.space.com/39240-when-to-see-planets-in-the-sky.html

Redd, N. T. (2017a, February). *How Did the Solar System Form?* Space.com; Space. https://www.space.com/35526-solar-system-formation.html

Redd, N. T. (2017b, May 5). *Asteroid Belt: Facts & Formation.* Space.com; Space.

https://www.space.com/16105-asteroid-belt.html

Redd, N. T. (2017c, December 9). *Olympus Mons: Giant Mountain of Mars.* Space.com; Space. https://www.space.com/20133-olympus-mons-giant-mountain-of-mars.html

Redd, N. T. (2018a, March 7). *How Fast Does Light Travel? | The Speed of Light.* Space.com; Space.

https://www.space.com/15830-light-speed.html

Redd, N. T. (2018b, March 28). *Red giant stars: Facts, definition & the future of the sun.* Space.com; Space. https://www.space.com/22471-red-giant-stars.html

Redd, N. T. (2018c, July 26). *What Is the Biggest Star?* Space.com; Space.

https://www.space.com/41290-biggest-star.html

Redd, N. T. (2018d, July 28). *What Is the Most Massive Star?* Space.com; Space.

https://www.space.com/41313-most-massive-star.html

Redd, N. T. (2019a, May 3). *Space Launch System: NASA's Next Generation Rocket.* Space.com;

Space. https://www.space.com/33908-space-launch-system.html

Redd, N. T. (2019b). News Feature: "Celestial snowman" starts to reveal its secrets. *Proceedings of the National Academy of Sciences, 116*(38), 18749–18752. https://doi.org/10.1073/pnas.1913629116

Rincon, P. (2014, October 16). *Mercury's hidden water-ice revealed.* BBC News; BBC News.

https://www.bbc.com/news/science-environment-29644406#:~:text=A%20Nasa%20spacecraft%20has%20provided,%2D%20could%20host%20water%2Dice.

Rocket HQ. (2020, September 4). *How Much Does It Cost You To Fill Up?* Rocket HQ; Rocket HQ.

https://www.rockethq.com/learn/personal-finances/how-much-does-it-cost-you-to-fill-up#:~:text=The%20amount%20of%20gas%20a,hold%2015%20or%2016%20gallons.

Ryden, B. (2021). *Lecture 34: Clusters and Superclusters.* Ohio-State.edu.

http://www.astronomy.ohio-state.edu/~ryden/ast162_8/notes34.html

Sci-News. (2020). *Astronomers Offer Solution to "Galactic Bar Paradox" | Astronomy | Sci-News.com.* Breaking Science News | Sci-News.com. http://www.sci-news.com/astronomy/solution-galactic-bar-paradox-08784.html

Sessions, L. (2020, December 17). *The Large Magellanic Cloud, our galactic neighbor | Astronomy Essentials | EarthSky.* EarthSky | Updates on Your Cosmos and World. https://earthsky.org/clusters-nebulae-galaxies/the-large-magellanic-cloud/

Sessions, L., & Gonzaga, S. (2021, April 7). *Alpha Centauri, star system closest to our sun | Brightest Stars | EarthSky.* EarthSky | Updates on Your Cosmos and World. https://earthsky.org/brightest-stars/alpha-centauri-is-the-nearest-bright-star/

Siegel, E. (2018, August 30). Our Motion Through Space Isn't A Vortex, But Something Far More Interesting. *Forbes.*

https://www.forbes.com/sites/startswithabang/2018/08/30/our-motion-through-space-isnt-a-vortex-but -something-far-more-interesting/

Silveira, J. (2015, December 22). *How big is the solar system? - Backwoods Home Magazine.* Backwoods Home Magazine. https://www.backwoodshome.com/how-big-is-the-solar-system/

Sky & Telescope. (2017, February). *Star-Trapping in Orion's Trapezium.* Sky & Telescope. https://skyandtelescope.org/observing/star-trapping-in-orions-trapezium/

Smith, K. N. (2016, October 28). *These are the spookiest, scariest storms on any known planet.*

Astronomy.com. https://astronomy.com/news/2016/10/weirdest-planetary-weather

Smithsonian Air and Space Museum. (2012). *Rovers - Exploring the Planets | National Air and Space Museum*. Si.edu. https://airandspace.si.edu/exhibitions/exploring-the-planets/online/tools/rovers.cfm

Smithsonian National Air and Space Museum. (2012). *Rovers - Exploring the Planets | National Air and Space Museum*. Si.edu. https://airandspace.si.edu/exhibitions/exploring-the-planets/online/tools/rovers.cfm

Solstation. (2014). *Gliese 229*. Solstation.com. http://www.solstation.com/stars/gl229.htm

Space.com Staff. (2017a, October 11). *Northern Lights: What Causes the Aurora Borealis & Where to See It*. Space.com; Space. https://www.space.com/15139-northern-lights-auroras-earth-facts-sdcmp.html

Space.com Staff. (2017b, October 28). *Dwarf Planets: Science & Facts About the Solar System's Smaller Worlds*. Space.com; Space. https://www.space.com/15216-dwarf-planets-facts-solar-system-sdcmp.html

SpaceX. (2020). *SpaceX*. SpaceX.

https://www.spacex.com/updates/crew-1-docks-to-iss/index.html

Sundermier, A. (2016, July 21). *Here are eight things you might not know about the moon*. Business Insider; Business Insider.

https://www.businessinsider.com/its-been-47-years-since-we-made-our-1st-footprints-on-the-moon-here -are-some-facts-you-might-not-know-about-it-2016-7#an-astronauts-footprint-can-last-a-million-years-o n-the-surface-of-the-moon-1

Technion. (2020, April 29). *The Origin of Arrokoth: Birth of a "Snowman" at the Edge of the Solar System*. SciTechDaily. https://scitechdaily.com/the-origin-of-arrokoth-birth-of-a-snowman-at-the-edge-of-the-solar-system/

The Associated Press. (2019, February 13). *NASA Mars rover Opportunity dies after exploring red planet for 15 years*. CNBC; CNBC. https://www.cnbc.com/2019/02/13/nasa-rover-opportunity-finally-dies-on-mars-after-15-years.html

The Editors of Encyclopaedia Britannica. (2021a). space shuttle | Names, Definition, Facts, &

History | Britannica. In *Encyclopædia Britannica*. https://www.britannica.com/technology/space-shuttle

The Editors of Encyclopaedia Britannica. (2021b). speed of light | Definition & Equation |

Britannica. In *Encyclopædia Britannica*. https://www.britannica.com/science/speed-of-light

The Nine Planets. (2019). *Solar System Facts | Information, Size, History and Definition*. The Nine Planets. https://nineplanets.org/solar-system/

The Planets. (2017, February 8). *Distances Between the Planets of the Solar System • The Planets*. The Planets. https://theplanets.org/distances-between-planets/

Tools of Cosmology. (2021). *The First Telescopes (Cosmology: Tools)*. Aip.org.

https://history.aip.org/exhibits/cosmology/tools/tools-first-telescopes.htm

UCL. (2021, June 13). *Dark matter is slowing the spin of the Milky Way's galactic bar*. UCL News.

https://www.ucl.ac.uk/news/2021/jun/dark-matter-slowing-spin-milky-ways-galactic-bar

University of Washington. (2021). *What Is Astrobiology? – uwastrobiology*. Washington.edu.

https://depts.washington.edu/astrobio/wordpress/about-us/what-is-astrobiology/#:~:text=Astrobiology %20is%20the%20study%20of,and%20stellar%20interactions%20and%20proce sses.

VideoFromSpace. (2021). SpaceX Crew-1's return to Earth - See the highlights [YouTube Video]. In *YouTube*. https://www.youtube.com/watch?v=R9CNRYEr674

Voosen, P. (2020, November 12). *Martian dust storms parch the planet by driving water into space*.

Science | AAAS. https://www.sciencemag.org/news/2020/11/martian-dust-storms-parch-planet-driving-water-space

Wall, M. (2020, February 13). *Arrokoth unmasked: NASA probe's "space snowman" encounter reveals how planetesimals are born.* Space.com; Space. https://www.space.com/new-horizons-arrokoth-flyby-planet-formation-clues.html

Williams, M. (2015, December 18). *What is the Big Bang Theory?* Phys.org; Phys.org.

https://phys.org/news/2015-12-big-theory.html

Williams, M. (2016, November 8). *What are Magellanic Clouds? - Universe Today.* Universe Today. https://www.universetoday.com/30537/what-are-magellanic-clouds/

Printed in Dunstable, United Kingdom